T0141819

Tailings Dam Management for the Twenty-First Century

Franco Oboni · Cesar Oboni

Tailings Dam Management for the Twenty-First Century

What Mining Companies Need to Know
and Do to Thrive in Our Complex World

With Contributions by Henry Brehaut

 Springer

Franco Oboni
Oboni Riskope Associates Inc.
Riskope
Vancouver, BC, Canada

Cesar Oboni
Oboni Riskope Associates Inc.
Riskope
Vancouver, BC, Canada

ISBN 978-3-030-19449-9   ISBN 978-3-030-19447-5   (eBook)
https://doi.org/10.1007/978-3-030-19447-5

This Springer imprint is published by the registered company Springer Nature Switzerland AG
The registered company address is: Gewerbestrasse 11, 6330 Cham, Switzerland

Preface

Recent dam failures have demonstrated the need for improvements in the design, management and regulation of tailings management facilities to reduce, and eventually eliminate, the risks of failure of existing and future facilities. The scope of the challenge is well documented in UNEP GRID-Arendal assessment report entitled "Mine Tailings Storage: Safety Is No Accident," which recommends that "Regulators, industry and communities should adopt a shared zero-failure objective to tailings storage facilities..." and identifies many areas where further improvements are required to meet this objective. In keeping with this objective, we believe that the application of cutting-edge risk assessment methodologies and risk management practices can contribute to the significant reduction and eventual elimination of dam failures through risk-based decision making and risk-informed decision making (respectively RBDM and RIDM).

Thus, the primary purpose of this book is to identify and describe those risk assessment approaches and risk management practices that must be implemented in order to develop a path forward to reach corporately, societally and ethically acceptable tailings dam risks. In support of this, we will describe appropriate, existing and proven risk management approaches. It will be shown that these can actually lead to the highest standards of tailings dam design and management required to reach UNEP goals and a sustainable and societally acceptable mining industry.

This book is based on over 30 years of worldwide consulting experience, relentless methodological research and development, prior books, courses, seminars delivered to corporate key stakeholders, public and governmental agencies, public hearings, expert witnessing in civil and criminal courts, and over 50 technical papers. Its purpose is to identify and describe those risk assessment approaches and risk management practices that must be implemented in order to develop a path forward to reach profitable and sustainable operations that will be corporately and societally acceptable from a risk standpoint. This will simultaneously cover the interests of lenders, investors and insurers as well.

This book is not a dam design book or manual; it is not a statistics and probabilities textbook; it does not aim at replacing MAC OMS manual; and finally, it is not a geotechnical engineering book/manual. Whenever details are needed, we provide references or summaries as appropriate, but the explanations given cover only what is needed for our discussion. All lists, examples are released as a mere illustration of a specific case and should never be construed as general indications or rules.

During the book design and writing phases, we kept asking ourselves who our readers would be, with the aim of custom tailoring the texts to their interests. It turns out the complexity of our modern society, and of the tailings systems and their governance, make it vain, in our minds, to focus on one single readership. Hence, we have attempted to write a single book that would deliver value to various categories in the mining world and the public. To do so, we have opted for various writing devices.

First, important comments and advice to readers are inserted as pertinent, making it possible to read the book as a "high-level" series of notes. Next, we have inserted as needed and pertinent, specific subsections entitled "Conclusions." These contain partial summaries related to the prior sections with our comments and discussions. Reading through these allows a "course-summary" type of reading, quickly grasping results and information while skipping the details of the discussions. Finally, interested readers can read the complete book text, gaining knowledge and becoming able to reference the various discussions.

Reading levels	Likely interested readership	Take-away and goals
High-level	"All mining public," Executive office	Why the industry needs a better approach to risk management and assessment. What is this better approach. How this better approach can be used to significantly improve the risk profile of tailings dams. Concepts consultants and middle management can use to convince head office that a better approach will add value
Course	Mid-management	The tailings manager's guide to a leadership approach to risk practice. Whether you are arriving in a new operation or you have managed a tailings facility for many years, this course will help you understand the evolution of codes, requirements and UNDP position. The course will help you identify fixes and enhancement to your risk approaches
Book	CEO/Key officers at Head Office	Deliver a clear understanding that it is time for a game-change (abandon common practice), describe what a better approach will look like, and illustrate how it should be used. Define questions you should be asking to Mid-Management and consultants in order to shed light on the design, construction, management and monitoring of tailings dams

Vancouver, Canada   Franco Oboni
                                                                                        Cesar Oboni

Acknowledgements

This book would not have been possible without the support of many people.

We very much want to thank Henry Brehaut who decided to assist us with enthusiasm, after ten minutes of conversation, when we met at Tailings and Mine Waste 2017. His perspectives on key management issues and research on recent accidents are greatly appreciated.

We are also grateful to our clients, who have pushed us to find new solutions for more and more complex problems. They are the ones who inspired us to break the boxes in which common practices have locked themselves in, triggering years and years of continuous research and development and field testing of our methodologies and procedures.

The authors wish to express their gratitude to Dr. Annett Büttner, Senior Publishing Editor, Springer Earth Sciences, Geography and Environment, and to Kim Williams, Kim Williams Books, for their timely and constructive criticisms.

The authors wish to express their love and gratitude to their beloved families, for their understanding and endless love, during the preparation of the manuscript.

About This Book

Recent dam failures have demonstrated the need for improvements in the design, management and regulation of tailings management facilities to reduce, and eventually eliminate, the risks of failure of existing and future facilities. The scope of the challenge is well documented in the United Nations Environment Programme (UNEP) report released in October 2017 in Geneva entitled UNEP GRID-Arendal assessment entitled "Mine Tailings Storage: Safety Is No Accident."

The UNEP report recommends that "Regulators, industry and communities should adopt a shared zero-failure objective to tailings storage facilities…" and has identified many areas where further improvements are required to meet this objective. In keeping with this objective, we believe that the application of cutting-edge risk assessment methodologies and risk management practices can contribute to the significant reduction and eventual elimination of dam failures through risk-informed decision making.

Thus, the primary purpose of this book is to identify and describe those risk assessment approaches and risk management practices that must be implemented in order to develop a path forward to reach societally acceptable tailings dam risks. In support of this, we will describe appropriate, existing and proven risk management approaches. It will be shown that these can actually lead to the highest standards of tailings dam design and management required to reach UNEP goals and a sustainable and societally acceptable mining industry.

This book discussion is split into three parts:

Part I, State of Affairs, looks at several examples of recent catastrophic failures from a forensic analysis point of view, as they were performed by various independent panels and entities. At the end of Part I, the various recommendations are distilled into one comprehensive "specs file" for the twenty-first century.

Part II, How to Build a Twenty-First-Century Comprehensive Risk Management Program, starts with a basic list of "DON'Ts," after which the tone turns positive and the text follows the logic of a real-life risk-informed decision-making approach. Indeed, each chapter corresponds to a necessary step to develop a modern risk assessment which can then lead to risk-informed decision making "from cradle to grave" of a tailings facility, or any other mining system.

Part III, Case Histories and a Look into the Future, features a risk-informed decision-making deployments example and a discussion on risk communication and concludes with a look into what we expect might happen in the years, decades to come.

Concrete examples and case histories support the discussion. In particular, we show:

1. How critical risks, and their necessary management controls, can be identified for both existing and new dams;
2. How cost-effective risk reduction strategies can be identified and included in the design and management of new projects;
3. How risk assessment approaches can provide a solid basis for the justification and prioritization of continuous improvement programs;
4. How an extant dam portfolio can be risk-prioritized in order to design mitigative measures while appropriately communicating risk to the public.

Contents

Authors and Contributor

About the Authors

Franco Oboni is a geotechnical engineer (Ph.D.) with over 35 years of experience and has specialized in quantitative risk assessment (QRA) for over 25 years. He leads Riskope, a Vancouver-based practice active internationally providing advice on quantitative risk assessment (QRA) and decision-making support. His clients include Global 1000 companies, large insurance companies, natural resource entities (mining companies, etc.), railroads, wharves, governments and suppliers. He consults on risk and crisis mitigation projects, risk and security audits and geo-environmental hazard mitigation studies on four continents. He has authored over 50 papers and has co-authored 2 books: *Improving Sustainability Through Reasonable Risk and Crisis Management* (2007) and *The Long Shadow of Human-Generated Geohazards: Risks and Crises* (2016). He delivers customized seminars (in English, French, Italian and Spanish) to industrial audiences worldwide. He was co-recipient of the Italian Canadian Chamber of Commerce (Canada West) 2010 Innovation Award.

Cesar Oboni has been involved in risk analyses and reviews for numerous facilities and organizations, communities, mining companies (coal mining, gold, copper and zinc mining), military bodies and transportation entities, including "at perpetuity" projects. In addition to his activities in Risk Assessment and Optimum Risk Estimates (ORE, the flagship product of Riskope for industries and projects around the world), he has been very active in the analysis of special and emerging risks, co-authoring a report on cyber-defense at national scale for an European country. Cesar analyzes and prioritizes residual risk of mining infrastructures, tailings and water treatment plants. His clients include Fortune 500 companies, large mining corporations, UN/UNDP, transportation and military bodies. He is the author of more than 20 papers published in the proceedings of international conferences and symposiums. He is the co-author two books: *Improving Sustainability Through Reasonable Risk and Crisis Management* (2007) and *The Long Shadow of*

Human-Generated Geohazards: Risks and Crises (2016). He was co-recipient of the 2010 ICCC, the Italian Canadian Chamber of Commerce (Canada West) Innovation Award.

Contributor

Henry Brehaut has spent his entire working life in the mining industry with experience in underground and open pit mining, mine and corporate development, environment and sustainable development and corporate management. As a corporate executive and director, he gained extensive experience in corporate governance and mine management, including chief operating responsibility for nine tailings dams. He currently acts as an independent consultant on sustainability issues as President of Global Sustainability Services Inc. and has provided executive and board level advice on sustainable development issues to governments, multilateral organizations, mining companies and industry associations. In 1999, he received the Prospectors & Developers Association of Canada Environmental Award for "helping to bring environmental and sustainable development issues to the forefront in Canadian and international mining." He has a B.Sc. in mining engineering from Queen's University and a MBA from the University of British Columbia.

Abbreviations

BAT	Best Available Techniques
CCF	Common Cause Failure
CDA	Canadian Dam Association
CEO	Chief Executive Officer
COO	Chief Operating Officer
COV	Coefficient of Variation
CRO	Chief Risk Officer
CRT	Corporate Risk Tolerance
CSR	Corporate Social Responsibility
DWR	Department of Water Resources
EIA	Environmental Impact Assessment
e-IDC	Extended IDC (or Extended Investigation, Design and Construction)
EOR	Engineer of Record
ESA	Effective Stress Analyses
ETA	Event Tree Analyses
FERC	Federal Energy Regulatory Commission
FoS	Factor of Safety
FTA	Failure Tree Analyses
GIS	Geographic Information System
H&S	Health and Safety
ICOLD	International Commission on Large Dams
IDC	Investigation, Design and Construction
IFT	Independent Forensic Team
InSAR	Interferometric Synthetic Aperture Radar
IoT	Internet of Things
IT	Information Technology
ITB	Independent Tailings Board
ITBR	Internal Tailings Review Board
IW	Intermittent Water Analysis
MAC	Mining Association of Canada

MCE   Maximum Credible Earthquake
MFL   Maximum Foreseeable Loss
ORE   Optimum Risk Estimates©
PAF   Perception Affecting Factors
PCM   Persistent Change Monitoring
PFMA   Potential Failure Modes Analyses
PGA   Peak Ground Acceleration
PIG   Probability Impact Graph
QPE   Qualified Professional Engineer
QPO   Quantitative Performance Objectives
QRA   Quantitative Risk Assessment
RBDM   Risk-Based Decision Making
RIDM   Risk-Informed Decision Making
SAR   Synthetic Aperture Radar
SLM   Silva–Lambe–Marr Method
SLO   Social License to Operate
TMS   Tailings Management System
TMW   Tailings and Mine Waste
TRF   Tailings Responsibility Framework
TSF   Tailings Storage Facility
UNEP   United Nations Environmental Protection
USA   Undrained Stress Analysis
USCOLD   United States Committee on Large Dams
USEPA   United States Environmental Protection Agency
WISE   World Information Service of Energy
WTP   Willingness to Pay

Chapter 1
Introduction

1.1 The Context

The future success of the mining industry is dependent on its ability to gain the trust of a wide range of stakeholders in order to obtain its social license to develop and operate (Social License to Operate (SLO) (https://www.riskope.com/2015/09/24/big-players-beware/)) mines while remaining competitive, profitable and fostering Corporate Social Responsibility (CSR) (https://www.riskope.com/2015/05/21/ethics-in-mining-oil-and-other-natural-resources-fields/). Our discussion of twenty-first-century tailings dams management starts with a dispassionate look at four major recent failures, revealing the gaps that will make it possible to develop twenty-first-century tailings dams management programs that are sustainable, reasonable, rational and add value to mining project and the industry as a whole.

While tailings management is only one of many major and oftentimes competing issues that must be addressed to ensure long-term success and sustainability, recent history has shown that the mining industry has yet to achieve the level of performance (https://www.riskope.com/2013/11/07/social-acceptability-criteria-winning-back-public-trust-require-drastic-overhaul-of-risk-assessments-common-practice/) and trust required to eliminate catastrophic dam failures and reach the industry's overarching goal. New methods—including dry paste, co-mingled, and whatever technology will be developed in the future—will not necessarily reduce risks. Indeed, in the last year or so, we have already reviewed new projects including some of these technologies that "take more risks" as the materials are deemed superior to standard deposition methods ones. This is a never-ending loop requiring the greatest attention to avoid future sorrow.

As a result, in October 2017 the United Nations Environment Program (UNEP) in Geneva released a UNEP-GRID Arendal assessment report entitled "Mine Tailings Storage: Safety Is No Accident" (Roche et al. 2017). The report urges States and the industry to *end deadly and damaging mining waste spills by adopting a zero-failure objective*.

© Springer Nature Switzerland AG 2020
F. Oboni and C. Oboni, *Tailings Dam Management for the Twenty-First Century*,
https://doi.org/10.1007/978-3-030-19447-5_1

We believe that reaching this objective requires significant improvement in many areas particularly through the application of sophisticated but sustainable quantitative risk assessment and risk management methodologies. These can contribute to the reduction and eventual elimination of dam failures through risk-based-decision-making and risk-informed decision making (respectively RBDM and RIDM).

Following the Federal Energy Regulatory Commission (FERC) Risk Guidelines (https://www.ferc.gov/industries/hydropower/safety/guidelines/ridm.asp) definition, Risk Informed Decision-Making (RIDM) applied to dams uses the likelihood of loading, dam fragility, and consequences of failure to estimate risk. Risk estimates are then used **to decide if dam safety investments are justified or warranted**.

For Risk-based decision making we follow the definition given by the Committee on Improving Risk Analysis Approaches (NRC US 2009). RBDM (http://www.au.af.mil/au/awc/awcgate/uscg/risk-qrg.pdf) organizes and orders information about potential nefarious events to **shore decision makers management choices**. RBDM asks questions like: What can go wrong? How likely are potential problems to occur? How severe might the consequences be? Is the risk tolerable? What can/should be done to mitigate the risk?

When comparing the two definitions it results that RBDM can be seen as a subset of RIDM, the two having a common trunk. In this book we refer to RBDM when an analysis shores and supports decision making whereas we consider RIDM when we discuss tactical and strategic risks, extending into formal financial analyses of mitigative alternatives.

We understand there are indeed companies, consultants and regulators around the world that have already a high level of commitment to tailings dam safety and have developed practices that have been instrumental in ensuring the safe operation of tailings dams under their watch in many countries (see Further Reading at the end of the book for some examples, among which are pleased to highlight the latest version of the MAC OMS manual (MAC OMS Guide 2019)). The challenge pointed out in the keynote lecture by Henry Brehaut at the Tailings Mine Waste 2017 Conference (Brehaut 2017) is to use this experience as the starting point in the development of standards, practices and guidelines that must be used by all companies to establish higher levels of performance, by governments for effective oversight of the industry and to provide the basis for a high level of public trust (https://www.riskope.com/2013/05/08/can-we-stop-misrepresenting-reality-to-the-public/) (Oboni et al. 2013).

"Trust us" no longer works; Demonstrated commitment supported by leading edge risk assessments will.

Higher levels of performance, standards and oversight require better definition of the quality designation for an industry practice. To this day this generally rests solely in the eye of the beholder, which is most likely to be the person or organization that has developed it. Common terms are good, best and, less often, leading practice. Brehaut pointed out that even the term "emerging practice" has recently been used to describe a practice that has been employed by some committed companies for over 20 years.

> A strategic intent is required to meet the needs of modern society and allow miners to pursue a sustainable future.

The keynote lecture added that:

> the concern with most of such designations stems from how they were developed. At the lowest level, it is a self-designation by an individual, company or an organization. Moving up the scale, a committee of experts, usually under the umbrella of a mining association, government or a professional body, is mandated to develop a set of practices or guidelines based on their collective experience. Such efforts require a consensus at the committee level and approval by the supporting organization. Because of varying degrees of commitment at both the committee and approval level, this inevitably leads to a less than aggressive pushing of the boundaries.

> Self-identified common practices and even so-called "best practices" have failed in preventing catastrophic failures.

If other stakeholders are involved, the resulting practices lead to the inclusion of more demanding and detailed expectations, such as more specific guidance and performance thresholds. The term "best practice" may be applicable to such situations. Leading practice should be forward-looking and defined as a practice that goes well beyond the norm, providing the highest degree of commitment and/or a significant improvement in risk reduction thus constituting a strategic intent (Hamel and Prahalad 2005).

The bottom line is that existing self-identified good, best and leading practices have failed and still fail to prevent catastrophic dam failures and significant progress is required. Furthermore, the more general the description of a practice, the easier it is for the less committed to claim they have adopted it. Therefore, this text aims at tightening the definitions and requirements with the goal of seeing a significant improvement in tailings management and risk reduction. Hence, through this book we will use a ISO 31000-compliant glossary (see Sect. 14.1).

> Twenty years ago (1997) Franco Oboni participated in a IUGS workshop in Honolulu aimed at a first attempt to define a glossary of risk-related technical terms specific to slopes and landslides. In short, the IUGS produced a glossary in "Quantitative Risk Assessment for Slopes and Landslides" (Working group on Landslides, Committee

on Risk Assessment, 1997). Actually we have kept that glossary alive, expanded it to be applicable to other fields of business and industries. It is available for everyone to download (https://www.riskope.com/knowledge-centre/tool-box/glossary/).

Using a well-defined glossary is paramount in any field, but especially so in the area of risk. As we will see, the use of fuzzy definitions has created significant confusion and has been identified as a cause of incidents and accidents. We strongly recommend readers to familiarize themselves with terms such as "risk" and "consequences", to cite only two examples, before going any further.

According to Eq. (1.1), risk R can be expressed, in its simplest form as the product of the probability of occurrence of an undesirable event P_f and the damages it potentially causes D (Kaplan and Garrick 1981; Lowrance 1976; Haimes 2009):

$$R = P_f \cdot D \qquad (1.1)$$

However, there is no universal consensus on the definition of risk, and in recent years the focus has shifted to cover the whole spectrum of probabilities, consequences and uncertainties (Aven 2012). In this book we will use $R = p_f * C$ where C represents the consequences (not necessarily only the damages).

As a matter of fact, both UNEP and Henry Brehaut have discussed qualitatively what mining companies need to know and do to achieve a significantly higher level of performance. They also presented a number of forward-looking leading practices necessary to significantly advance the level of care required by the mining industry. Both present important principles and leading practice goals but *no methodological suggestions*.

Further along this line of discussion, in a recent lecture Prof. Morgenstern stated (Morgenstern 2018) stated that the dominant cause of dams' failures arises from deficiencies in engineering practice associated with the spectrum of activities embraced by design, construction, quality control, quality assurance, and related matters. It is no surprise that the absence (or inadequacy) of peer review is considered a limiting factor in reducing the incidence of tailings facility failures (Morgenstern 2010; Caldwell 2011). However, in the aftermath of the recent Corrego do Feijao Dam failure near Brumadinho in Brazil (2019) we read in the media coverage of the event that a third-party inspection occurred and that, as in many other historic tailings dams failures, the dam apparently showed no signs of significant distress shortly before the failure. This indicates that peer review and traditional monitoring and inspections cannot be seen as universal panacea. Limited engineer involvement and an alarming disconnect arising from the poor definition of potential consequences of mishaps and their societal ripple effects—an aspect indeed mostly ignored in codes—leave professionals ample room for biases and censoring applied to potential losses (Oboni et al. 2013; CDA 2014). At the other end of the spectrum lie excessively prescriptive codes that hinder professional judgment and good engineering. Furthermore, inadequate understanding of undrained failure mechanisms leading to static liquefaction with extreme consequences is a factor in about 50% of the cases studied by Morgenstern (Morgenstern 2018). *Here again, no methodological suggestions were offered.*

1.2 The Objective of Twenty-First-Century Tailings Dams Management

This book builds on the leading practices suggestions cited above and focuses on the identification of those advanced quantitative risk assessment (QRA) and risk management methodologies required for their implementation and rational risk-informed decision making.

> The methodologies required for the implementation of the leading practice suggestions exist and deliver.

The methodologies exist, have delivered proven results over the last two decades and overcome well known pitfalls and documented problems with common practice risks assessments (Cox et al. 2005; Cox 2008; Cresswell nd; Hubbard 2009; Chapman and Ward 2011; Oboni and Oboni 2012; Thomas et al. 2014). All players need to up their game and the latest round of incremental improvements should be looked at as only the first step of a committed process leading to the highest level of performance possible.

As stated earlier, this book focuses on "how to" related to the principles and leading practices formulated by UNEP and Henry Brehaut. Unless necessary for the completeness of discussion it leaves aside themes that require not methodological interventions but administrative procedures. Among these we can cite: permitting, boards composition, professional qualifications, codes and guidelines, etc. Due to obvious space limitations, we will focus on the design and operating stages of the mining cycle.

It is during the design stage that the probabilities of the consequences of failure need to be evaluated in relation to the proposed deposition method and site location. It is during the operating stage that mining companies must strive for the lowest reasonable and sustainable combination of likelihood of failure and consequences (that is, risk), and hence develop rational and sustainable risk mitigation.

> The hazard landscape must be clearly defined in order to avoid misunderstandings and confusion.

The terms "lowest", "reasonable" and "sustainable" must be discussed. Indeed, as zero risk does not exist, and risk is a function of two parameters—i.e., the likelihood of failure p_f and related consequences C—it may well be that rational and sustainable risk mitigation be the result of a pair (p_f, C) where possibly only one is at its lowest value, or together they represent indeed the best possible compromise. Thus, it is at that stage that risks have to be compared with corporate and societal tolerances. A risk below tolerance is defined as tolerable, whereas a risk above tolerance is defined as intolerable. Later (Chap. 13) we will also discuss terms such as acceptable and societal.

Finally, while the design process will address recognized natural hazards, it must be acknowledged that additional design risks may exist due to gaps in the knowledge base supporting current technical standards leading to a hazard not being recognized. Variations in the degree of professional experience, judgment and conduct may also be factors.

> We define as hazard scenarios any malfunctioning (or deviations from the intended level of performance) of the system or any of its elements "as is". System "as is" means with the present level of mitigation and controls and with the quality of investigations, design and maintenance which becomes apparent during the preparation of the study. Thus design choices such as Factors of Safety (FoS), and the effects of length and number/density of reconnaissance boreholes are included in the probability of failure of each element as described later, but do not constitute an hazard as defined above. The same occurs for management, maintenance and monitoring. However deviations from the intended level of care in management, maintenance and monitoring are considered to be hazards.

Bad management can be considered a hazard like another and should be included in any risk approach. However, as we will discuss later, the hiccups of "business as usual" of any kind are not to be considered as hazards or risks in order to avoid paralysis by analysis. Uncertainties are discussed as needed through this book (see also Sect. 7.4).

Management risks can arise because of low corporate commitment, economic feasibility pressures, and insufficient resources provided to support dam design and the implementation of management systems. Regulatory risks may be introduced as part of the permit approval process and through inadequate compliance and enforcement activities.

Another way of looking at how a tailings facility can be exposed to further risks is to consider the dynamics within each major participant. The term "regulatory capture" has been used by the Auditor General of British Columbia (BC AG 2016) to describe the situation where the regulator, created to act in the public interest, may instead, in certain situations, serve the interests of a company or the industry. Using the same perspective for mining companies, "economic capture" could be described as the situation where short- or long-term economic factors are given precedence over sustainability commitments and permit obligations. For the geotechnical community, "client capture" could be used to describe situations where the engineer may be influenced by client pressures in the performance of their work.

1.3 The Plan

The twenty-first-century tailings dams management discussion is split into three parts.

Part I: State of Affairs

Chapter 2 describes the recent scene. Two catastrophic tailings dams failures (Mount Polley and Samarco) are reviewed based on Independent Review Panel and other forensic analyses.
Chapter 3 is similar to Chap. 2, but looks at recent hydro-dams failures and major incidents.
Chapter 4 looks at the "statistics" of one hundred years of major tailings dams failures and shows how a systemic approach can be used to describe the failure process of tailings dams.
Chapter 5 analyzes public reactions and "what the public wants".
Chapter 6 draws the conclusions from the documents of the previous chapters and shows, under the form of "specifications", why new approaches are indeed needed for twenty-first-century tailings dams management and what such approaches should entail.

Part II: How to Build a Twenty-First-Century Comprehensive Risk Management Program

Chapter 7 explains what not to do under the form of a list of DON'Ts.

> You can follow step by step the build-up of a modern risk management program from Chap. 8, System definition, all the way to Chap. 14, Risk assessment for the twenty-first century.

Chapter 8, System definition, shows how to dissect a mining system into macro-elements, that is, elements which are amenable to analyses.
Chapter 9 discusses hazard identification.

Chapter 10, Defining probabilities of events, discusses how probabilities of single events can be evaluated and updated, and delivers some examples of portfolio analyses.

Chapter 11 focuses attention on the most frequent failure mode of dams—instability—and delves into the details of its analysis.

Chapter 12, Defining consequences, explains how to consider the multidimensional aspect of hazard consequences.

Chapter 13 discusses risk tolerance from approaches both historic and modern.

Chapter 14, Risk Assessment for the twenty-first century, delves into the methodological details of modern risk assessments in compliance with the "specs" defined in Part I, and in particular Chap. 6.

Part III: Case Histories and a Look into the Future

Chapter 15, Risk-Informed Decision-Making, is a set of case histories used as deployment examples.

> Chapter 15 features a series of Risk-Informed Decision-Making case histories, deployments examples.

Chapter 16, One last word, includes some important notes on what the reader should be able to perform using this book.

Finally, **Chapter 17**, Path forward, defines what the possible evolution of world-class procedures should be.

References

Aven T (2012) The risk concept - historical and recent development trends. Reliab. Eng. Syst. Saf. 99: 33–44

[BC AG 2016] Auditor General of British Columbia (2016) An Audit of Compliance and Enforcement of the Mining Sector http://www.bcauditor.com/pubs/2016/audit-compliance-and-enforcement-mining-sector

Brehaut H (2017) Catastrophic dam failures path forward, Keynote lecture, Tailings and Mine Waste 2017, Banff, Nov 5–9, 2017

Caldwell J (2011) Slimes dam - aka tailings storage facility - failure and what it meant to my mining mindset, cited in Chambers & Higman (2011), p. 18, n. 24

[CDA 2014] Canadian Dam Association (2014) Technical Bulletin, Application of dam safety guidelines to mining dams

Chapman C, Ward S (2011) The Probability-impact grid - a tool that needs scrapping, in: How to manage Project Opportunity and Risk, 3rd ed., ch. 2, pp 49–51, Chichester, GB, John Wiley & Sons

Cox, LA Jr (2008) What's Wrong with Risk Matrices? Risk Analysis 28(2): 497–512

Cox LA Jr, Babayev D, Huber W (2005) Some limitations of qualitative risk rating systems, Risk Analysis 25(3): 651–662

Cresswell S (nd) Qualitative Risk & Probability Impact Graphs: Time For A Rethink? http://www.intorisk.co.uk/resources/white-papers/qualitative-risk-assessmens

Haimes YY (2009) On the Complex Definition of Risk: A Systems-Based Approach. Risk Analysis 29: 1647–1654

Hamel G, Prahalad CK (2005) Strategic Intent, Harvard Business Review, Jul-Aug 2005, https://hbr.org/2005/07/strategic-intent

Hubbard D (2009) Worse than Useless. The most popular risk assessment method and Why it doesn't work, in: The Failure of Risk Management. Ch. 7, pp. 117–144, Hoboken, Wiley & Sons

Kaplan S, Garrick BJ (1981) On the quantitative definition of risk. Risk Analysis 1(1):11–27

Lowrance W (1976) Of Acceptable Risk - Science and the Science and the Determination of Safety, Technical report, Los Altos, CA, William Kaufmann Inc.

[MAC OMS Guide 2019] Mining Association of Canada (2019) Developing an Operation, Maintenance, and Surveillance Manual for Tailings and Water Management Facilities SECOND EDITION, Feb. 2019

Morgenstern NR (2010) Improving the safety of mine waste impoundments. In: Tailings and Mine Waste '10, Vail, CO. October 17–20, 2010, Boca Raton: CRC Press

Morgenstern NR (2018) Geotechnical Risk, Regulation, And Public Policy, The Sixth Victor de Mello Lecture, 2018, SOILS and ROCKS An International Journal of Geotechnical and Geoenvironmental Engineering I, 41(2): 107–129

[NRC US 2009] National Research Council (US) Committee on Improving Risk Analysis Approaches Used by the U.S. Science and Decisions. Advancing Risk Assessment, EPA, Washington (DC), National Academies Press (US) (http://www.nap.edu/)

Oboni C, Oboni F (2012) Is it true that PIGs Fly when Evaluating Risks of Tailings Management Systems? Mining 2012, Keystone CO

[Oboni et al. 2013] Oboni F, Oboni C, Zabolotniuk S (2013) Can We Stop Misrepresenting Reality to the Public?, CIM 2013, Toronto. https://www.riskope.com/wp-content/uploads/Can-We-Stop-Misrepresenting-Reality-to-the-Public.pdf

Roche, C., Thygesen, K., Baker, E. (eds). (2017). Mine Tailings Storage: Safety Is No Accident. A UNEP Rapid Response Assessment. United Nations Environment Programme and GRID-Arendal, Nairobi and Arendal, ISBN: 978-82-7701-170-7 http://www.grida.no/publications/383

Thomas P, Bratvold RB, Bickel JE (2014) The Risk of Using Risk Matrices, SPE, Economics & Management 6(2): 56–66

Part I
State of Affairs

Part I, State of Affairs, looks at several examples of recent catastrophic failures from a forensic analysis point of view, as they were performed by various independent panels and entities. At the end of Part I, the various recommendations are distilled into one comprehensive "specs file" for the twenty-first century.

Chapter 2
Two Recent Catastrophic Tailings Dams Accidents

This chapter sets the recent scene. Two catastrophic tailings dams failures (Mount Polley and Samarco) are reviewed based on Independent Review Panel and other post-mortem analyses.

2.1 Mount Polley

On 4 August 2014, a breach occurred within the Perimeter Embankment of the tailings storage facility (TSF) at the Mount Polley copper and gold mine in south-central British Columbia. The breach (Fig. 2.1) resulted in the release of an estimated 25 million cubic metres of wastewater and tailings that resulted in surface environmental damage and the transport of tailings and water into a nearby lake. No lives were lost but remediation costs are estimated by the company to be in the order of 70M$US (the estimate reached 205M$US in 2018), and the company's market value was adversely affected.

The impact of the dam failure was judged by most external observers to be catastrophic. If measured against the current consequence characterization criteria established by the government of British Columbia, the consequence classification would be at the significant level, three levels below the extreme impact classification. From a corporate perspective, the impacts of the dam failure would have to be judged as being of the highest order based on considerations related to the loss of market value, operating earnings and public trust as well as the costs of remediation. In corporate terms, such impacts would be judged as being material, requiring the highest level of risk management and corporate governance.

In order to identify the cause of the failure and to assist in the identification of improvements for tailings management in general, the initial response by the government of British Columbia was to appoint an Independent Expert Engineering

© Springer Nature Switzerland AG 2020
F. Oboni and C. Oboni, *Tailings Dam Management for the Twenty-First Century*,
https://doi.org/10.1007/978-3-030-19447-5_2

Fig. 2.1 Google Earth view of the Mount Polley mine tailings pond before (top) and after (bottom) the dam break of August 4th 2014

Investigation and Review Panel to report on the breach as described in Sect. 2.1.1 below. In direct response to the Panel Report, the Association of Professional Engineers and Geoscientists of B.C. prepared "The Professional Practice Guidelines – Site Characterization for Dam Foundations in B.C." (APEGBC 2016) to address the recommendation in the Panel that guidelines be prepared that would lead to improved site characterization. As a parallel activity the Auditor General of British Columbia issued a report titled "An Audit of Compliance and Enforcement of the Mining Sector" (BC AG 2016) with its main recommendations pertaining to risk identification and management being described in Sect. 2.1.2. The government of British Columbia also implemented changes to the Health, Safety and Reclamation Code for Mines in British Columbia, primarily on the basis of the above documents.

2.1.1   Mount Polley Panel Report

The government of British Columbia established an independent expert engineering investigation and review panel (Panel 2015) to investigate and report on the Mount Polley breach. The primary purpose of the Panel was to investigate and report on the cause of the failure of the tailings storage facility, including the identification of any mechanism(s) of failure. The Panel was also asked to identify or comment on what actions could have been taken to prevent the failure, any technical, management or other practices that may have enabled or contributed to the mechanism(s) of failure, and to identify any changes that could be considered to reduce the potential for future such occurrences.

The Panel concluded: "The breach of the Perimeter Embankment on 4 August 2014 was caused by shear failure of dam foundation materials when the loading imposed by the dam exceeded the capacity of these materials to sustain it". The Panel further stated: "The dominant contribution to the failure resides in the design" in that the design did not take into account the complexity of the sub-glacial and pre-glacial geological environment associated with the Perimeter Embankment foundation. In this regard it was also stated: "The type and extent of pre-failure site investigations were not sufficient to detect this stratum or identify its critical nature".

In making the above statements the Panel noted that:

> … in conducting its inquiry, the Panel limited itself to relying on interviews and on the documents that it received from the various stakeholders, which were sufficient to determine root cause of the breach. The Panel did not conduct its process according to formal legal procedures. To do so would have extended the length of this investigation and would have entered into an assessment of roles and responsibilities, which is beyond the Panel's authorization. As a result, the Panel is not able to offer an adequate assessment of the role of management and oversight in its contribution to the cause of the failure. In particular, the Panel has not explored the relationship between the designers and owner, contractual or otherwise. Accordingly, the Panel is unable to ascertain the circumstances that contributed to key decisions.

However, in keeping with its terms of reference the Panel was able to offer valuable insights by identifying the hazards that were unique to the moment of the failure and commenting on those aspects of the design, management and government processes that allowed those causal or contributing factors exist in the first place.

Design References

With regard to design, the Panel was of the opinion that:

- "There were ambiguities in the governing factor of safety, adapted from Canadian Dam Association (CDA) Guidelines never intended for tailings dams. An FoS = 1.3 design criterion using peak effective-stress strength left little margin for error, and trigger-level factors of safety for critical piezometric conditions were even lower at FoS = 1.1."
- "Looking specifically at the failure, it was deemed desirable to increase the target FoS to 1.5 since the TSF was operating more or less continually at full capacity."

- "Mount Polley illustrates that dam safety guidelines intended to be protective of public safety, environmental and cultural values cannot presume that the designer will act correctly in every case. To do so defeats the purpose of FoS criteria as a safety net. In this, the CDA Guidelines are unable to achieve their intended purpose. Neither is the Province well served, to the extent that the Ministry of Energy and Mines (MEM) has incorporated compliance with these guidelines as a statutory requirement."
- Had a planned buttress "... been in place... failure would have been prevented."
- "Had the downstream slope in recent years been flattened to 2.0 horizontal to 1.0 vertical, as proposed in the original design, failure would have been avoided."
- "The Observational Method was adopted as a design philosophy, but misapplied. For reasons not unrelated to planning shortcomings, instrumentation was relied upon to substitute for definitive input parameters and design projections. The Mount Polley dam was ill-suited to this approach, for both practical and strategic reasons. The steep slopes and constant construction activity on the Perimeter Embankment prevented installation of instruments at optimal locations. More importantly, the instrumentation program was incapable of detecting critical conditions because, once again, the critical materials and their critical mode of undrained behavior were not recognized."
- "The Panel found it disconcerting that, notwithstanding the large number of experienced geotechnical engineers associated with the Mount Polley TSF, the overall adequacy of the site investigation and characterization of ground conditions beneath the Perimeter Embankment went unquestioned. This may reflect a regional issue, or possibly one of wider extent. Regardless, it calls for a concerted effort to improve professional practice in this area."

Management References

With regard to management practices, the Panel observed or was of the opinion that:

- Mine planning: "A lack of foresight in planning for dam raising contributed to the failure. Successfully executing the raising plan required intimate coordination of impoundment water-level projections, production and transport of mine waste for raising, and seasonal constraints on construction. This made the tailings dam contingent at the same time on the water balance, the Mine plan, and the weather. But instead of projecting these interactions into the future, they were evaluated a year at a time, with dam raising often bordering on ad hoc and only responding to events as they occurred. The effects were twofold: a near over-topping failure in May 2014, and restrictions on mine waste availability that produced the over-steepened slopes and deferred buttress expansion."
- Water balances: The water balances used did not "...provide a reliable approach to establishing adequate capacity for tailings and water storage" and "uncertainties in the water balance input parameters combined with uncertainties in climatic conditions and construction schedules cannot provide a robust design for water containment".
- Water impoundment levels: High impoundment water levels were a major cause of chronic problems in maintaining a tailings beach around the perimeter of the

dam. At the breach section, water was in direct contact with the upstream zone of tailings fill when failure occurred. This increased the piezometric level in the upstream zone above what it would have been had a wide tailings beach been present. The Panel's analyses show that this had some influence on dam stability, although it was not the dominant factor.

- Water impoundment volume. The panel was of the opinion that:

  1. "Had the water level been even a meter lower and the tailings beach commensurately wider, this last link might have held until dawn the next morning, allowing timely intervention and potentially turning a fatal condition into something survivable."
  2. "It was water erosion that transported the bulk of the tailings, and these fluvial processes ended when the supply of water was exhausted. Had there been less water to sustain them, the proportion of the tailings released from the TSF would have been less than the one-third that was actually lost."
  3. "It is not clear to the Panel why it took so long to design and implement a water treatment strategy that would provide for a significant reduction in the amount of surplus water stored on the TSF."

- Downstream dam slope: The adoption of the steeper 1.3H:1V downstream slope may have been due to limited material availability or other aspects related to mine planning.
- Other concerns related to management practices:

  1. Departures from intended designs related to filter and transition construction.
  2. Failure of much of "the as-placed filter material...to meet applicable filter criteria and requirements for internal stability of its grading."
  3. Chronic problems with maintaining the tailings beach continued.
  4. Related absence of a well-developed tailings beach violated the fundamental premise of the design as a tailings dam, not a water-storage dam.
  5. Even if not contributing directly to the failure, some design details were problematic. Already thin to begin with, reducing the core width from 8 to 5 m made it even more vulnerable to differential settlement and cracking. Both the filter and transition zones were just 1 m wide, placing great demands on their performance.

Panel Recommendations

In response to its mandate to identify any changes that could be considered to reduce the potential for future such occurrences, the Panel's recommendations pertaining to risk identification and management are summarized as follows:

1. Future permit applications for a new TSF should be based on a bankable feasibility that would have considered all technical, environmental, social and economic aspects of the project in sufficient detail to support an investment decision, which might have an accuracy of ±10–15%. More explicitly, it should contain the following:

    a. A detailed evaluation of all potential failure modes and a management scheme for all residual risk.

    b. Detailed cost/benefit analyses of best available techniques (BAT) tailings and closure options so that economic effects can be understood, recognizing that the results of the cost/benefit analyses should not supersede BAT safety considerations.

    c. A detailed declaration of quantitative performance objectives (QPOs).

2. Corporations proposing to operate a TSF should be required to be a member of the Mining Association of Canada (MAC) or be obliged to commit to an equivalent program for tailings management, including the audit function.

3. The utilization of independent tailings review boards should be utilized to enhance validation of safety and regulation of all phases of a TSF.

4. The concept of QPOs to improve regulator evaluation of ongoing facilities should be utilized to strengthen regulatory oversight.

    The panel also stated:

5. "In the view of the Panel, the fundamental need is to improve the geological, geo-morphological, hydro-geological and possibly seismotectonic understanding of sites proposed for tailings dams in B.C. This improved understanding should account for the likely scale associated with variability so that site investigations can be planned with enhanced reliability."

6. With regard to industry standards (CDA):

    a. "Recognizing the limitations of the current Canadian Dam Association (CDA) Guidelines incorporated as a statutory requirement, develop improved guidelines that are tailored to the conditions encountered with TSFs in British Columbia … that emphasize protecting public safety."

    b. "The Panel anticipates that this will result in more prescriptive requirements for site investigation, failure mode recognition, selection of design properties, and specification of factors of safety."

7. Government Role:

    The Panel, while being impressed with the skill and commitment of regulatory staff, identified the need to strengthen the current regulatory operations in British Columbia. Their main recommendation was to introduce the concept of critical control measures QPOs as an aid to improved oversight. However, in Chap. 8, "Regulatory Oversight", the Panel noted that the regulator had "limited ability" to influence design issues having to rely on the expertise and professionalism of the Engineer of Record (EOR). It was further noted that "The relationship between the Regulator and the EOR can result in different opinions being expressed that are not easy to resolve without independent input."

2.1.2 *Mount Polley Auditor General's Report*

In May 2016, the Auditor General of British Columbia issued report titled "An Audit of Compliance and Enforcement of the Mining Sector" (BC AG 2016). The stated purpose of the audit was to determine whether the regulatory compliance and enforcement activities of the Government of British Columbia pertaining to the mining sector were protecting the province from significant environmental risks. The audit was initiated prior to the Mount Polley breach to assess regulatory effectiveness at all operating mines in British Columbia but was later expanded to include a major section focusing on major issues related to this specific incident.

The overall audit finding was that "…compliance and enforcement activities of the mining sector are inadequate to protect the province from significant environmental risks." The findings that support such a conclusion and the resulting recommendations are well documented in the report and, because of this, this section will only identify and discuss matters pertaining to the use of risk assessment methodologies in support of improvements in regulatory oversight and industry performance.

One of the important concepts introduced in the audit was that not only should the public be protected from significant environmental risks but the public should also be informed about such risks. One of its recommendations is that government should:

> …publicly disclose its rationale for granting a permit…. Specifically, information should include how factors such as economic, environmental, and social attributes were considered in the determination of public interest.

The report also identifies the need for a risk-based approach to compliance verification activities with inspections based on identified risks.

2.2 Samarco Tailings Dam Breach

On November 5, 2015, Samarco's Fundão dam failed. It was one of Samarco's two primary tailings dams at their mine site in Brazil. This failure led to a significant volume of mine tailings being released (Fig. 2.2) with the result that three communities were flooded and a number of other communities further downstream were also affected.

Nineteen people died; five community members and fourteen people who were working on the dam facility at the time of the failure. Approximately 700 people lost their homes, seven bridges were destroyed, and access roads and 100 km of fencing were damaged. Over 2000 ha of riverside vegetation and agricultural land were impacted, affecting the lives of about 7000 families.

Fig. 2.2   Samarco's Fundão dam failure flood. *Photo* Senado Federal

2.2.1   External Investigation Review Panel

Samarco and its owners, BHP Billiton Brasil and Vale, jointly engaged a law firm to coordinate an external investigation into the immediate cause of the breach of the Fundão tailings dam with the stated purpose of providing a detailed technical understanding of the immediate cause of this failure. To carry out this work, the Fundão Tailings Review Panel (Fundão Panel 2016) was formed, composed of four geotechnical engineers who were specialists in water and tailings dams "…to provide its independent and unbiased professional judgment and expertise in determining the immediate cause(s) of the incident."

"The Panel systematically identified and evaluated multiple causation hypotheses. It further imposed hypothesis testing by means of the following three questions that the candidate failure mechanism was to be able to explain:

1. Why did a flowslide occur?
2. Why did the flowslide occur where it did?
3. Why did the flowslide occur when it did?

Forensic methods adopted by the Panel integrated multiple lines of evidence: observations from eyewitness accounts; data and imagery in Geographic Information System (GIS) format; field evidence from subsurface exploration by the Panel and others; advanced laboratory testing; and sophisticated computer modelling. Responding to the above three questions for hypothesis testing demanded a high level of quantification and exhaustive detail in each of these aspects of the Investigation's evidence-based approach."

The Fundão Panel was asked not to assign fault or responsibility to any person or party and because of this it was not able to provide insights as to how the underlying hazards went unrecognized or were allowed to exist at the time of failure.

2.2.2 Cause of Failure

As reported by the Fundão Panel in its report, "Fundão Tailings Dam Review Panel: Report on the Immediate Causes of the Failure of the Fundão Dam" (Fundão Report), the original design was based on employing an unsaturated sand zone for basic dam support. It was further noted that unsaturated sand is not amenable to liquefaction and because of this the original design was robust in this regard. In summary, the dam failed by liquefaction flowsliding as a consequence of a chain of events and conditions.

The Fundão Panel further reported that:

> a change in design brought about an increase in saturation which introduced the potential for liquefaction. As a result of various developments, soft slimes encroached into unintended areas on the left abutment of the dam and the embankment alignment was set back from its originally-planned location. As a result of this setback, slimes existed beneath the embankment and were subjected to the loading its raising imposed. This initiated a mechanism of extrusion of the slimes and pulling apart of the sands as the embankment height increased. With only a small additional increment of loading produced by the earthquakes, the triggering of liquefaction was accelerated and the flowslide initiated.

The Fundão Panel considered the "evolutionary character of the design and operation to be extraordinarily complex" and employed a structured approach to the examination of a large number of potential failure modes. Its conclusions were that:

1. "The surviving liquefaction trigger mechanism is static load increase …with its two subsidiary processes: undrained shearing and deformation-related extrusion. Both of these might be operative either with or without cyclic pore pressure contribution from the November 5, 2015 earthquake series."
2. "Also important… are the antecedent events and conditions… that allowed or promoted static liquefaction at the left abutment. These are: (1) saturation of the sand; (2) water encroachment that allowed slimes deposition on the tailings beach; (3) the alignment setback; and (4) the increased height of the setback resulting from continued raising of the dam."

References

[APEGBC 2016] Association of Professional Engineers and Geoscientists of British Columbia (2016). Professional Practice Guidelines – Site Characterization for Dam Foundations in BC. V1.0. https://www.apeg.bc.ca/For-Members/Professional-Practice/Professional-Practice-Guidelines

[BC AG 2016] Auditor General of British Columbia (2016) An Audit of Compliance and Enforcement of the Mining Sector http://www.bcauditor.com/pubs/2016/audit-compliance-and-enforcement-mining-sector

[Fundão panel 2016] Fundão Tailings Dam Review Panel (2016), Report on the immediate Causes of the Failure of the Fundao Dam. Panel: NR Morgenstern, SG Vick, CB Viotti, BD Watts

[Panel 2015] Independent Expert Engineering Investigation and Review Panel (2015) Mount Polley Report. Report on Mount Polley Tailings Storage Facility Breach. Province of British Columbia.https://www.mountpolleyreviewpanel.ca/final-report

Chapter 3
Examples of Recent Catastrophic Hydro-Dam Accidents

3.1 Oroville Dam Spillway Accident

This chapter is similar to Chap. 2, but looks at recent hydro-dams failures and major incidents.

The Oroville Dam is an earth-fill embankment water dam in the Sierra Nevada foothills northeast of San Francisco owned and operated by the California Department of Water Resources (DWR). At 770 ft (235 m) high, it is the tallest dam in the United States and serves for water supply, hydroelectricity generation and flood control. The dam impounds Lake Oroville and has a capacity to store more than 4400 million cubic metres of water. We selected the Oroville case because of its interest, although Oroville dam is not a tailings dam and the dams itself was not breached.

3.1.1 What Happened

In February 2017, heavy rainfalls required the release of water through the main flood control spillway. On 7 February 2017, erosion of a large area of concrete and underlying rock in the main spillway was observed (Fig. 3.1) and discharge rates were reduced to protect the integrity of the structure. This resulted in water levels behind the dam rising, due to further heavy rainfalls, to the point that the water began to be discharged over the crest of the emergency spillway. The emergency spillway discharge channelized as it flowed across the natural terrain downstream of the crest structure causing extensive erosion, with some of the erosion areas head-cutting aggressively toward the emergency spillway crest structure. At this point, on 11 February, the authorities triggered the evacuation of more than 180,000 people downstream of Lake Oroville. Two days after the evacuation order, when water levels

© Springer Nature Switzerland AG 2020
F. Oboni and C. Oboni, *Tailings Dam Management for the Twenty-First Century*,
https://doi.org/10.1007/978-3-030-19447-5_3

Fig. 3.1 Erosion of a large area of concrete and underlying rock in the main Oroville Dam spillway. *Photo* Dale Kolke/California Department of Water Resources

behind the dam had stabilized, the authorities downgraded the order to an evacuation warning and gradually allowed residents to return to their homes.

While the main embankment dam did not fail, its integrity was significantly threatened, as indicated by the decision to evacuate 180,000 people, which would not have been taken lightly. The consequences of 4400 million cubic metres of water being released are hard to imagine. Referring to the spillway problems as an incident does not properly express the seriousness of what happened and the need to use it as a learning opportunity.

In response to this incident, the Federal Energy Regulatory Commission required the DWR to engage an Independent Forensic Team (IFT) to develop findings and opinions on the causes of the incident. Two leading dam engineering and dam safety associations in the United States were asked by DWR to recommend a team of six members, all of whom were accepted DWR and subsequently engaged. The terms of reference adopted by the IFT are as follows:

> To complete a thorough review of available information to develop findings and opinions on the chain of conditions, actions, and inactions that caused the damage to the service spillway and emergency spillway, and why opportunities for intervention in the chain of conditions, actions, or inactions may not have been realized. Evaluations of actions, inactions, and decisions for the various stages of the project (pre-design, design, construction, operations, and maintenance) will consider the states of practice applicable to the various time periods involved.

A summary of its key findings, with particular attention to risk identification and risk informed decision-making are presented in what follows.

3.1.2 The IFT Report

Summary

The "Independent Forensic Team Report—Oroville Dam Spillway Incident, January 5, 2018" (IFT Report 2018) provides an excellent analysis of not only the geotechnical physical aspects of the incident but also of the human, organizational and industry factors that created the hazards that existed at the time of the incident. In the introductory summary of the report it stated that "…the incident was caused by a complex interaction of relatively common physical, human, organizational, and industry factors, starting with the design of the project and continuing until the incident".

The main physical factors related to the failure of the spillways systems are described as being related to "unrecognized inherently weak design and as-built conditions affecting the initial failure area" and "unrecognized poor foundation conditions incompatible with design." It was noted that flowing water resulted in uplift forces beneath the service spillway chute slab that exposed the underlying poor quality foundation rock to unexpected severe erosion, resulting in removal of additional slab sections and more erosion. For the emergency spillway, the report states that the principal physical factor contributing to the damage was clearly the presence of significant depths of erodible soil and rock in features orientated in a manner that allowed rapid erosion toward the crest control structure. Details can be found in Chap. 4, "The Physics of What Happened", of the IFT Report.

The question of why and how the owner, regulator and industry practices failed to prevent the incident forms a very valuable part of the IFT's deliberations. The human, organizational and industry factors that led to important risks not being recognized and addressed are described in Chap. 5, "Why the Incident Happened" and Chap. 6, "General Organizational, Regulatory, and Industry Factors". The following sections in this chapter will build on the findings of the IFT Report Summary and Chap. 7, "Lessons To Be Learned," as they apply to risk identification and risk management from an industry perspective.

What the IFT Learned

While not stated as a lesson to be learned, the most important observation of the IFT team was that human judgments, decisions, actions and inactions can be significant contributing factors in creating the conditions that allow a system failure to occur. Understanding a failure or incident from the standpoint of the physical factors and mechanisms that were involved is a necessary first step but the IFT believes that it is only the starting point of a process that must lead to an analysis of why those factors or mechanisms were allowed to exist in the first place. It is for this reason that their lessons to be learned focus more on management commitment and the decisions required to translate the highest standards into practice. The IFT notes further that their lessons, while developed from their work on the Oroville water dam incident, will apply generally to industry level practices. With regard to this present book, we believe them to be equally applicable to industry practices for tailings dams.

Potential Failure Modes Analyses (PFMAs): The IFT recognizes that the PFMA process is a very useful tool in helping designers and companies identify and manage their risks. However, based on the Oroville incident investigation they believe the current PFMA process:

- can encounter difficulties in properly characterizing hazards and related risks for large or complex systems, including accounting for human and operational aspects in failures;
- can demonstrate a tendency to oversimplify complex failure modes involving multiple interactions of system components by defining failure modes as a linear chain of events;
- may not explicitly consider how broader organizational factors, such as culture and decision-making authority and practices, can contribute to failure.

This lesson stems from the findings that the PFMA process, related to the service spillway, failed to identify or adequately examine existing maps and borehole logs that showed the weak quality of the strongly weathered rock in the area, did not recognize the significant level of hazard associated with the spillways and, as a result, did not have sufficient information for the development of adequate warning signs and trigger mechanisms. More specific insights and recommended improvements related to PFMAs can be found in Appendix F3 of the IFT Report.

Comprehensive Facility Reviews: This lesson derives from the finding that the comprehensive facility reviews had failed to identify and, therefore, understand the physical factors that led to failure of the spill way chute. The IFT Report notes that without this information, the development of potential failure modes and the application of the PFMA process was compromised. It was further stated that the Oroville reviews

> ... have tended to focus on what has changed since the prior review, based on observations from inspection, surveillance, monitoring, and operations during the prior five-year interval, rather than more comprehensively reviewing long-term performance history and comparing the original design and construction with current best practices for design and construction.

The focus of the lesson regarding comprehensive facility reviews is on the adequacy of the design and management plans in terms of judging a chance for failure or unsatisfactory performance, the adequacy of information to make such judgments and, if not, the need for further study or evaluation. The IFT is also of the opinion that the reviews should be:

- thorough, taking advantage of all available information;
- critical and independent, rather than relying largely on the findings of past reviews;
- completed by people with appropriate technical expertise, experience, and qualifications to cover all aspects of design, construction, maintenance, repair, and failure modes of the assets under consideration.

Physical Inspections: The roots of this lesson were found in the use of visual inspections as the primary part of the risk management program structured to monitor the integrity of the water retaining facility on an ongoing basis. The IFT identified the need for owners and operators to be better informed to identify risks and manage dam integrity by expanding the inspection program to include supporting information provided by subsurface and non-destructive testing, ongoing facility reviews and risk analysis methodologies.

Regulatory Compliance: The lesson here is very simple. It is that "Compliance with regulatory requirements is not sufficient to manage dam owners' and public risk." The IFT recognizes that regulators play an essential role in the management of dam facilities but notes that their efforts are typically narrowly focused and restricted by available resources. Despite such limitations the IFT noted that DWR had relied too heavily on the regulators and the regulatory process.

Dam Safety Program and Risk Management: The IFT is generally of the view that a dam owner should have: (1) a strong commitment to dam safety based on strong top-down leadership; (2) a comprehensive risk management framework with risk-informed dam safety decision-making as an integral part; (3) sufficient staff and funding to identify and manage dam safety issues on a proactive basis, rather than merely struggling to keep up with regulatory requirements on a reactive basis. While this lesson was directed at DWR, it is included here because of its general applicability.

3.2 The Xe-Pian Dam Disaster

In this section we briefly discuss another hydro-dam failure and draw some links to the tailings dams world portfolio.

A series of events starting Sunday, 22 July 2018, led to a Laotian hydro-dam collapse with catastrophic consequences on the population downstream (Fig. 3.2). The collapsed dam was part of the Xe-Pian Xe-Namnoy hydroelectric power project, then still under construction. The project is located in the Attapeu Province in central Laos and involves Laotian, Thai and South Korean firms. Lao PDR is particularly rich in vertical drops with hydro-generating potential. As a matter of fact, the country had forty-six operational hydroelectric power plants in 2017 with a further fifty-four under construction.

The feasibility study for this particular project was reportedly completed in November 2008, construction began in February 2013 and commercial operations were expected to begin in 2019 (Poindexter 2015). The project includes two main dams (Xe-Pian and Xe-Namnoy) and five subsidiary dams. The dam that collapsed was the "Saddle Dam D", one of the subsidiary dams. The structure was reportedly 8 m wide (at the crown), 770 m long and 16 m high. The purpose of the subsidiary dams of this project is to "help divert water around a local reservoir", as reported by the owner and delivered by local media.

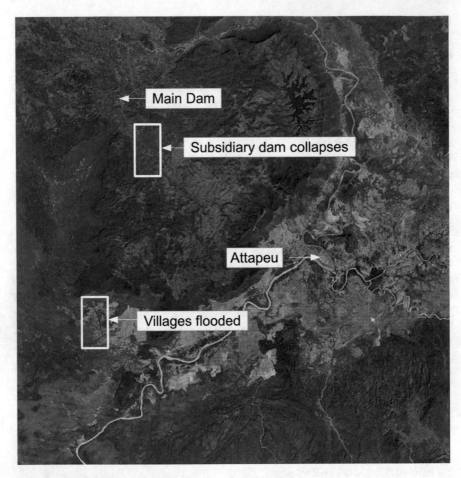

Fig. 3.2 Xe-Pian dam disaster area. Satellite view from Google Earth

After the Oroville dams problems in 2017 (Sect. 3.1), this case encompasses a catastrophic and devastating failure of a "minor" hydro structure. Are we seeing the development of a new trend in risk, thanks to climate change? Last year, a dam on the Nam Ao River, also under construction, failed as well. In that case, however, no deaths were reported.

3.2.1 Meteorological Conditions

According to public news (Sosnowski 2018) broadcast in February 2013, the Philippines, Vietnam, Cambodia and Laos were going to face the greatest threats from tropical storms and typhoons, including flooding rainfall and/or damaging winds and

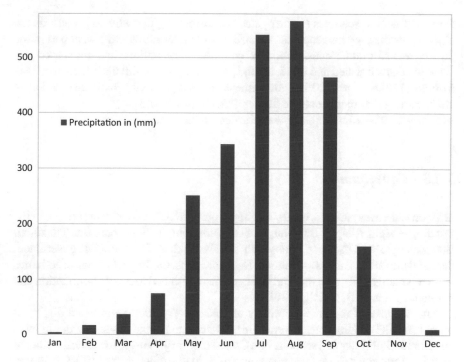

Fig. 3.3 Attapeu average precipitations

seas. Figure 3.3 shows Attapeu average monthly precipitations (https://en.climate-data.org/asia/laos/attapeu-1818/). Precipitation in the region is the lowest in January, with an average of 3 mm. Most precipitation falls in August, with an average of 566 mm. Tropical Storm Son Tinh on July 18–19, 2018 caused flooding in various provinces. Media reported heavy rainfall in Attapeu Province resulted in dangerously high river and dam levels (UN OCHA 2018).

3.2.2   Dam Collapse Chronology

The main Thai stakeholder reportedly declared that the dam "was fractured" after "continuous rainstorm(s)" caused a "high volume of water to flow into the project's reservoir". The timeline was approximately as follows:

Sunday 21:00 local time (14:00 GMT)—The dam is found to be partially damaged. The authorities are alerted and villagers near the dam start to be evacuated. A team is sent to repair the dam, but is hampered by heavy rain, which has also damaged many roads. This constitutes a sad textbook example of external inter-dependency.
Monday 03:00—Water is discharged from one of the main dams (Xe-Namnoy dam) to try to lower water levels in the subsidiary dam. We do not know whether both

dams had flood discharges (weirs? gates? pennstocks?), nor why only one was the object of emergency manoeuvres. We also do not know at this stage what part of the project was still under construction, as commercial operations had not started yet. Were the dams finished and filled, but the generating plants still under construction?
Monday 12:00—The Lao PDR government orders villagers downstream to evacuate after learning that there could be further damage to the dam.
Monday 18:00—More damage is confirmed at the dam.

3.2.3   Consequences

The consequences are unfortunately of tragic proportion. The accident resulted in the flooding of eight villages: Ban Mai; Ban HinLath; Ban ThaSengchan; Ban Thahintai; Ban Sanong; Ban Thae; Ban Phonsa-ath; Ban Nongkhae. The Lao PDR government declared the affected areas as National Disaster Areas. On July 24 response activities were reportedly still underway and UN agencies continued to gather information and assess the impacts (UN OCHA 2018).

The collapse of the dam reportedly affected nearly 7000 people and displaced more than 1000. Many houses were damaged, forcing people to seek shelter in local government buildings and schools. Red Cross teams in the Attapeu branch distributed clothing, food and drinking water to households in the affected areas (IFRC/Lao Red Cross 2018).

3.2.4   Risk-Related Concepts

As we have done in the aftermath of other catastrophic failures, we will not discuss responsibilities or faults but focus instead on a few risk-related concepts.

Lao PDR is hit by frequent tropical storms and cyclones.[1] We do not know if there is a climate change trend developing at this time, but these events are far from a surprise in Lao PDR. Tropical cyclones are not direct hazards, since their force is normally diminished (reaching tropical storm level) once they reach Lao PDR from the South China Sea. However, they can produce flooding as a consequence of heavy rainfall. Up to three cyclones hit the country annually (UNISDR et al. 2018).

We do not know how many failures there might have been in the country over the course of its history, nor how many dam-years[2] the country actually has. If we consider a round number of 50 hydro projects, at this time we had 1/50 failures. Combining this with a frequency of major tropical storm of, let's say, 1/10, we have $p_f = 1/10/50 = 1/500$ (or 0.002, see horizontal line in Fig. 3.4). If we look at historical hydro-dams failure rate of 10^{-4}–10^{-5} (1/10,000–1/100,000) and

---

[1] The difference between tropical storm and cyclone lies in the wind speed.

[2] Normally expressed as the product of the number of dams by the years in existence.

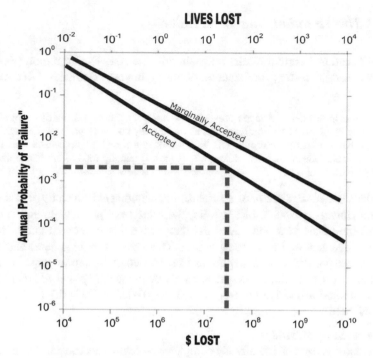

Fig. 3.4 Whitman's tolerance thresholds (Whitman 1984). In the '80s it was assumed a "cost of casualty" would be 1M USD as seen by comparing the top scale with the bottom scale. See Sect. 13.1.2 for a discussion on this subject

even 10^{-6} (1/1,000,000) (C. Oboni and F. Oboni 2012) the jump up to 1/500 is frighteningly significant. With $p_f = 0.002$ and damages in the tens of M\$ (vertical line in Fig. 3.4) the newly developed dams would clearly be bordering societal acceptability even without considering harm to people. Chap. 13 discusses in detail risk tolerance thresholds.

We do not know any dam-specific design peculiarities—e.g., what the return flood was, design criteria, etc.—so we will avoid any speculation. However, what seems obvious at this time is that meteorological alert system would have helped and maintaining access to the operation would have a positive impact. The question is: will we see a significant policy change in Lao PDR in the aftermath of this catastrophe? That seem to be the case in most countries hit by dire events of this magnitude.

3.2.5   The Situation One Week Later

In an international current-affairs magazine for the Asia-Pacific region (Fawthrop 2018) we read interesting confirmation of our preliminary estimates. We quote verbatim:

> The Lao dam disaster in Attapeu province has cast a long, dark shadow of doubt about the safety standards and viability of dozens of other hydro-power projects. This entirely preventable man-made catastrophe left 6000 people homeless from floods and over 1000 villagers unaccounted for… The Lao government had initially tried to lay blame on the heavy monsoon rain.

Media reported the director of Southeast Asian Studies at University of Wisconsin-Madison University, stated, like we did, the rains were predictable and "normal" for this time of the year and added the dam broke due to a combination of poor water management and poor construction. This conclusion is general and worldwide, as we pointed out in a publication entitled A systemic look at tailings dams failure process (http://www.riskope.com/wp-content/uploads/2016/10/Paper-17_A-systemic-look-at-TD-failure-TMW-2016_07-11_revised.pdf) (F. Oboni and C. Oboni 2016).

A General State of Affairs

The Laotian government has already completed 51 hydro-electric dams with another 46 under construction. Most of these projects have begun in the last 10 years, with the government arguing that hydro-power could lift landlocked Laos out of poverty while becoming "the battery of Asia." Based on prior experience with earth dams, we all know that even the "simplest" dam can generate significant risks. For example, agricultural dams in the US created a national problem decades ago, before they were systematically breached to reduce their risks.

China has a long history of dam disasters, including the worst dam disaster in history, which killed 171,000 and displaced 11 million inhabitants in 1975. Tailings dams in the mining industry and dykes and levees built to protect cities and agricultural land have a long history of failures, as witnessed by historic records, for example, in the Netherlands. Starting with geotechnical investigations, design, construction, monitoring and management all the way to end of service life, dam construction is indeed a highly challenging, potentially high-risk process.

A Public Reaction is Brewing

In the words of the director of Southeast Asian Studies at University of Wisconsin-Madison University, Iain Baird:

> Many Lao people are very upset about the breaking of the dam, calling for the Lao government to re-assess plans to build so many dams in the Mekong River Basin. This level of concern is unprecedented in Laos.

Long before the failed dam was built, the director co-authored an Environmental Impact Assessment (EIA) on Fisheries (Baird 2011) with the following recommendations:

Considering its important and relatively intact natural resources in fishes as well as in forests, mammals, birds and other organisms, we recommend that plans to dam and divert the upper Xe Pian be dropped from the Xe Nam Noy-Xe Pian project. Only in this way can the fisheries, wildlife and village resources of the area be protected and preserved for future generations.

Sadly, the recommendation was not followed-up. Lao Prime Minister Thongloun Sisoulith admitted it was the worst disaster faced by the small Southeast Asian country in decades.

Closing Remarks

In this failure case and similarly to tailings dams portfolios, we can conclude that the results of decisions and choices made at the inception of dam projects will generate risks for decades in the future. If our first estimate is indeed correct, we will hear more serious news coming from that corner of the world, in some cases widely passing national borders.

References

Baird I (2011) The Don Sahong Dam: Potential Impacts on Regional Fish Migrations, Livelihoods and Human Health. Critical Asian Studies. 43. 211–235 https://doi.org/10.1080/14672715.2011. 570567

Fawthrop T (2018) Laos' Dam Disaster May Not Be Its Last. *The Diplomat*, 2 August 2018 https:// thediplomat.com/2018/08/laos-dam-disaster-may-not-be-its-last/

[IFRC/Lao Red Cross 2018] International Federation of Red Cross And Red Crescent Societies (2018) Lao dam collapse floods nearby towns and affects thousands of people https://reliefweb.int/report/lao-peoples-democratic-republic/lao-dam-collapse-floods-nearby-towns-and-affects-thousands

[IFT Report 2018] Independent Forensic Team Report, Oroville Dam Spillway Incident, January 5, 2018 https://damsafety.org/sites/default/files/files/Independent%20Forensic%20Team%20Report%20Final%2001-05-18.pdf

Oboni C, Oboni F (2012) Is it true that PIGs Fly when Evaluating Risks of Tailings Management Systems? Mining 2012, Keystone CO

Oboni F, Oboni C (2016) A systemic look at tailings dams failure process, Tailings and Mine Waste 2016, Keystone CO, USA, October 2–5, 2016 https://www.riskope.com/wp-content/uploads/2016/10/Paper-17_A-systemic-look-at-TD-failure-TMW-2016_07-11_revised.pdf

Poindexter GB (2015) Project Update: 410-MW Xe-Pian Xe-Namnoy hydroelectric project in Laos, Hydro-Review, 19 November 2015 https://www.hydroworld.com/articles/2015/11/us-1-02-billion-410-mw-xe-pian-xe-namnoy-hydroelectric-project-includes-3-dams.html

Sosnowski A (2018) 2018 Asia spring forecast: Tropical storms to threaten Philippines to eastern India; Drought may build from eastern Turkey to Kazakhstan. AccuWeather (https://www.accuweather.com/en/weather-news/asia-spring-forecast-2018-tropicalstorms-to-threaten-philippines-to-eastern-india-while-drought-may-build-from-easternturkey-to-kazakhstan/70004154), February 26, 2018

[UNISDR et al. 2018] United Nations Office for Disaster Risk Reduction et al. (2018) Country Assessment Report for Lao PDR: Strengthening of Hydrometeorological Services in Southeast Asia https://www.unisdr.org/files/33988_countryassessmentreportlaopdr%5B1%5D.pdf

[UN OCHA 2018] UN Office for the Coordination of Humanitarian Affairs (2018) Lao PDR: Flooding - Office of the UN Resident Coordinator Situation Report No. 01 (as of 24 July 2018) https://reliefweb.int/report/lao-peoples-democratic-republic/lao-pdr-flooding-office-un-resident-coordinator-situation

Whitman RV (1984) Evaluating calculated risk in geotechnical engineering. J. Geot. Engineering 110(2): 145–188

Chapter 4
Historic Failures "Statistics"

4.1 A Hundred Years of Tailings Dams Lesson Learned

This chapter looks at one hundred years of tailings dams major failures "statistics" and shows how a systemic approach can be used to describe failure process of tailings dams.

Past and future probabilistic quantitative failure behaviour of tailings dams has been studied and published respectively in 2013, 2014 and 2015 at the Tailings and Mine Waste conferences (Oboni and Oboni 2013; Oboni et al. 2014; Caldwell et al. 2015). The aim was to provide quantitative measures to the predictive performances and various mitigation measures and levels of hazards/risk of tailings dams. In this section we summarize the findings of those papers.

At the conference Tailings and Mine Waste (TMW) 2013 (Oboni and Oboni 2013) we attempted the first estimate of the rate of failure of major tailings dams and compared their risks to human life to well-known societal tolerance. After stating the limitations of the available data and the lack of a clear definition of what constituted a major failure in commonly available statistics, we found rates varying between 10^{-3} (the decade around 1979) to 2×10^{-4} (the decade around 1999) major failures per dam year as shown in Table 4.1.

Table 4.1 Approximate pf of tailings dams based on preliminary sets of historic data (Oboni and Oboni 2013)

Where	When (decade)	p_f	Approx p_f
World-wide	Around 1997	$44/(3500 \times 10)$	10^{-3}
World-wide	Around 1999	$7/35,000$	2×10^{-4}
US	Around 1979 and Around 1999	7 or $8/(1000 \times 10)$	7 or 8×10^{-4}

© Springer Nature Switzerland AG 2020
F. Oboni and C. Oboni, *Tailings Dam Management for the Twenty-First Century*,
https://doi.org/10.1007/978-3-030-19447-5_4

In the paper of the following year (Oboni et al. 2014) we showed quantitatively how, over time, multiple hazards hits would significantly increase the probability of failure of a dam and lead to intolerable future risks. The paper concluded:

Especially in the case of TDs located in areas where demographic pressure leads to settlements in the downstream areas, social and legal consequences of a failure will dramatically increase. This will particularly be the case if the methodologies used to perform the risk assessments prove to be in disconnect with the needs of our modern society.

Finally, at TMW 2015 (Caldwell et al. 2015) we were delighted to examine the result of a new study of tailings dam historic failures (Bowker and Chambers 2015) which used detailed data and actuarial techniques to define the historic rate of failures of tailings dams after attempting to define what constitutes "serious" and "very serious" tailings dams failures. Cumulating those dimensions, we stated:

The common practice approach of using oversimplified consequence functions (with "or" clauses between consequences dimensions) is often used in research papers because of scope/budget limitations, but should not be accepted for a rational world-wide approach to decision making and tailings risks management for an industry that has significant societal impacts like mining. Tailings accidents generate multiple direct and indirect consequences on the environmental, human, health and safety (H&S), operational and reputational areas and we believe it is time for the mining industry as a whole to adopt a uniform consequence function.

Our paper concluded that:

It is comforting that the results of the "quick and dirty" 2013 (Oboni and Oboni 2013) study reached globally comparable results to the 2015 (Bowker and Chambers 2015) very deep and solid analytical approach. We noted that the selection of the time frame has a large influence on the conclusions of the 2015 study and therefore we recommended these comparative studies to be performed with constant duration (for example decade by decade) to avoid the hazard of drawing misleading conclusions. "Averaging" over 70 years, during which so many conditions have changed, may indeed mask decennial spikes. To prove this it is enough to note that the accident rates have actually decreased by 15–24% from the 1990–1999 to the 2000–2009 decades using the 2015 study's own data.

Finally it was stated that:

The aim of zero tailings failures is impossible to achieve (with the present world dams' portfolio). Tailings dams will continue to fail. In fact, in the long term all tailings facilities will spiral toward significant increases of their probability of failure and, when they fail, the tailings will run toward downstream rivers, lakes, and the ocean as they did at every failure to date. We demonstrated that consequences are not necessarily correlated, in one way or another, with dam height or pond volume. As in many industries the "scary stuff" is not necessarily the riskier one. Our practice and research have shown that the probability of failure is, or will be, often way higher in smaller structures than in major ones, simply because more care is taken for larger structures than for "insignificant" ones.

Well documented examples such as the 1974 failure of the Bafokeng dam in Rustenburg, SA, or the 1985 Val di Stava Dam collapse in Northern Italy are there to show that "extreme" consequences can actually occur. We also demonstrated that the rate of fatalities in the tailings "industry" lies way above the generally accepted "safe" thresholds for hazardous industries. The number of existing, operational, and

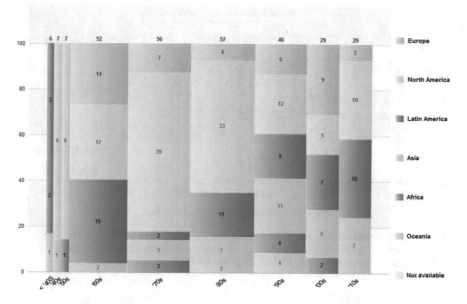

Fig. 4.1 Graphic representation of dams failures (https://www.riskope.com/2017/02/22/hundred-years-lessons-learned-tailings-dams-failures/). Columns bear decade's accidents (total at the top of column; column width is proportional to the total), colours correspond to geographic location, numbers in the cells correspond to number of accidents in related geographic area for that decade

closed TSFs around the world makes it necessary to prioritize the mitigation tasks, if we want to achieve a higher quality, be it at corporate or at national levels."

Figure 4.1 shows a summary of the full publicly available record of hundred years of dams failures (https://www.riskope.com/2017/02/22/hundred-years-lessons-learned-tailings-dams-failures/) (originally from Bowker and Chamber 2015). Given the nature of tailings dams, their construction time and expected service life and closure, the effects of today's risk mitigation programs will only slowly become visible because the world portfolio will contain mitigated and unmitigated (legacy) dams. During that period the public will perceive at best a status quo and the industry credibility and SLO will remain at stake (Oboni and Oboni 2014a, b; Oboni et al. 2013). This is why the October 2017 UNEP-GRID Arendal (Roche et al. 2017) report (See Sect. 4.1) recommends establishing an accessible public-interest, global database of mine sites, TSFs and related research.

In summary, thanks to the historic studies performed to date it is possible to assume the a priori range of the annual probability of failure of a dam goes from 10^{-5} to 10^{-3} based on 100 years performance history. The smaller value—i.e., 10^{-5}—would be for "top of class" structures.

For dams portfolios, if the design FoS are not known, prior analyses not available the "indirect hybrid assessment method" can be used. This approach is also based on historic approach to failure records (Oboni and Oboni 2013), empirical observations (Oboni et al. 2014) and a systemic failure model (Oboni and Oboni 2016b) which uses

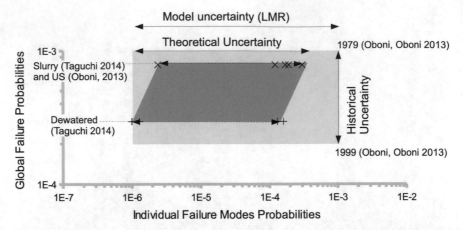

Fig. 4.2 A priori probabilities of failure for different tailings systems (slurry and dewatered) with their basic level of uncertainties. The vertical axis displays the global failure probability, whereas the horizontal axis shows the failure probabilities from individual failure modes

causality analysis in relation to design, construction, inspections, and maintenance. These papers have shown it is possible to establish a priori a range for a portfolio with very little actual specific data. By applying this type of procedure perceived lack of precision is compensated for with efficiency and preliminary understanding of ranges. Values of the probability of failure of specific failure modes for slurry and dewatered tailings dams have been derived and compared (Taguchi 2014) to the initial estimates (Oboni and Oboni 2013), as depicted in Fig. 4.2.

We can expect that the "world portfolio" of reasonably designed, built, maintained and managed dams lies within the rectangle bound by model uncertainty and historic uncertainty as already cited in the prior section. That means that we can expect that the world portfolio of reasonably designed, built, maintained and managed dams has a annual probability of failure between 2×10^{-4} and 10^{-3}, although the individual failure modes may have probabilities in the range of 10^{-4}–10^{-6}. We also see that the purely theoretical approach by Taguchi shows that dewatered tailings would tend to bring the probability of failure towards the bottom of the historical range, provided, of course the dewatering is effective and does not generate excessive risk taking based on its promises.

External symptoms such as beach width, wet spots, good or bad maintenance and deformations will modify the a priori p_f determined above. The same occurs if poor foundation conditions are known or suspected. A poorly designed or built or maintained/managed dam would lie outside of the rectangle, and reach failure by a ascending trajectory towards the left.

Please note that Fig. 4.2 shows the limit value of 10^{-6} for some dewatered tailings individual failure modes, which coincides with a value commonly considered the credibility threshold in evaluations.

4.2   A Systemic View on Tailings Dams Failure Processes

In this section we present a systemic approach of the "failure chain process" during the service life of dams (Oboni and Oboni 2016b). Investigations, design and construction (IDC) and then management, monitoring and water balance control (extended IDC, or e-IDC) of the dam are analysed with a probabilistic causality analysis based on publicly available incident and accidents data from the last hundred years. The predictive model, geared toward filling the gap between common practice and the "path to zero failures" goal (Mount Polley Report 2015), accommodates data-mining analytics. The model "constructs" the probability of failure of a dam which is consistent with factual historical world-data. The causality of various factors entering in the e-IDC process and other elements in the dam's service life can then be individually discussed with a sensitivity analysis. We then show where and how e-IDC process mitigative actions can benefit the most, with practical example. Special attention is spent on common cause failure (CCF) in operations, risk assessment and peer reviewing and inspections of tailings dams. CCF means that a "hidden" shared cause may lead parallel components to fail, annihilating the theoretical redundancy they have (or were designed for) (Mahesh 2014).

As we will see below, historic records show seepage and internal erosion are less frequent failure triggers than stability and seismic loading, likely due to historic state-of-the-art knowledge and understanding. Thus we have decided to focus the attention on those in this book. Interestingly, and for the same state-of-the-art reasons, liquefaction does not appear as a cause of failure in "statistical studies" published before the mid-1990s (we discuss liquefaction in Hazard Identification, see Sect. 9.3).

4.2.1   Design Process and Common Practice Risk Assessment

After each failure the mining community sees codes evolve and imposes tougher criteria and dam's specifications (TSM Task Force 2015). Factors of Safety (FoS) are eventually increased by empirical consensus among experts, whereas risk assessment methods have generally remained unchanged over almost half a century.

The relation between FoS and the probability of failure is often misunderstood, together with the multidimensional nature of potential damages to the environment, infrastructure and human beings. Codes that allow designers to use, for example, FoS $= 1$ under some pseudo-static seismic condition (CDA 2014, their Tables 3–5) actually accept that a tailings dam undergoing that seismic event would have the same chance of surviving/failing than a coin toss ($p_f = 0.5$). If the considered quake had a probability of 1/100 then the estimate of the risk under seismic loading would be pfs $= 0.5/100 = 5 \times 10^{-3}$ times the consequences of the failure (annually), respectively 10^{-3} times the consequences, if the quake has a probability of 1/500 annually. These are certainly not safe conditions with respect to public expectation or published tolerance thresholds (Oboni and Oboni 2013). Failure Mode and Effects Analyses (FMEAs) (https://www.riskope.com/2015/01/15/failure-modes-and-effects-analysis-fmea-risk-methodology) and Probability Impact Graphs (PIGs) remain the common practice risk assessment methodology (Oboni and Oboni 2012) despite their know limitation and misleading aspects (Chapman and Ward 2011; Cox 2008; FAA 2002; Hubbard 2009; NASA 2007). FMEAs lack the finesse needed to evaluate or predict the suggested progress "toward zero failures". Furthermore FMEAs do not help bringing any significant conclusion when comparing alternatives, cannot measure the efficiency of the (potential) mitigative measures implementation and compare them against themselves or even just determine if they are sufficient.

Finally, a significant number of risk studies we review do not start with a tailings system definition (See Chap. 8), its functional analysis and they confuse hazards, risks and consequences (Oboni et al. 2016) leading to misleading results. It is for example rather common to see "insufficient FoS" considered as a hazard (or a risk), whereas such deficiencies are generally the result of deliberate human choices (excessive audacity, errors and omissions, insufficient efforts). We will take a rather extreme, but logical, line of thinking, stating that dams failures find, in the vast majority, their root-causes in human choices and not in natural events.

At the centre of this reasoning there is the concept of credibility threshold we introduced at the end of the prior section. Many industries consider the limit of credibility at around $1/100,000–1/1,000,000$ $(10^{-5}–10^{-6})$ (Comar 1987; Wilson and Crouch 1982; Renshaw 1990), so it can be stated that any event above that limit is not an "Act of God" (or, following modern times buzz-words a "Black Swan") and should therefore be foreseen and planned for. We will also note that, reportedly, the vast majority of dam failures has occurred for other causes than "Black Swan" natural events, but again for "chains" of gradual deviances, which become "normalized" over time, stemming from investigations, design, construction, management and long-term monitoring.

4.2.2 Future Performance of the World-Wide Portfolio

As stated earlier, given the nature of the structures under consideration, their construction time and expected service life and closure, the effects of today's risk mitigation programs will only slowly become visible because the world portfolio will contain mitigated and unmitigated (legacy) dams. During that period the public will perceive at best a status-quo and the industry credibility and SLO will remain open to scrutiny and criticisms (Oboni and Oboni 2014a, b; Oboni et al. 2013).

It will be very difficult to evaluate progress. Factors such as climate change, seismicity (again, not necessarily "Black Swans"), increase in population and environmental awareness (consequence side of the risk equation) will further complicate the situation. Thus public outcry and hostility toward the mining industry, fuelled by the diffusion of information and communication technology, will likely increase and lead to prosecutions and larger fines. Due to the same influencing factors negligence and Force Majeure implications will certainly drastically change in the coming decades.

4.2.3   Tailings Dams Failure Processes

The elements described above show the need for a systemic approach of the "failure chain process" through investigations, design and construction (IDC) of tailings dams and then service-life management and long-term monitoring (e-IDC).

For the discussion we opted for a probabilistic causality analysis. Publicly available incident and accidents data from the last hundred years were again used. The predictive model is geared toward filling the gap between common practice and "path to zero failures" goal and accommodates data-mining analytic and future "lesson learned" that could make it possible to perform Bayesian updates (Dezfuli et al. 2009) after the first estimates (after the a priori estimate) (see Sect. 10.1.2).

The model has to include CCF (Stott et al. 2010) in operations, risk assessment, peer reviewing and inspections of tailings dams, at least in a simplified way, for the sake of completeness and explicit inclusion of conflict of interest and complacency (Oboni et al. 2013).

The Reliability Model
Engineering structures (and machinery, but also processes, including e-IDC processes) are systems consisting of a number of structural/physiological elements that can individually fail. The way elements are connected and their reliability X_j, where $X_j = 1 - p_{fj}$ define the reliability of the whole system (Eqs. 4.1, 4.2). For a series system (Eq. 4.1), failure of an element results in failure of the whole system. Reliability of the system is the product of the reliability of its elements. Equivalently, the system fails if any component fails.

Success:

$$\bar{X} = \prod_1^N \bar{X}_j \qquad (4.1)$$

A parallel system (Eq. 4.2) is a redundant system that is successful, if at least one of its elements is successful.
Success:

$$\bar{X} = 1 - \prod_1^N \left(1 - \bar{X}_j\right) = \coprod_1^N \bar{X}_j \qquad (4.2)$$

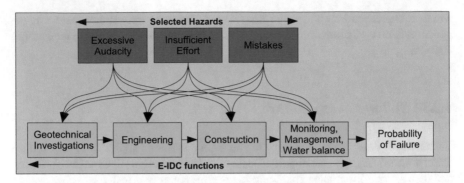

Fig. 4.3 Functional scheme of the e-IDC with the hazards selected for this study

In Fig. 4.3 we identify four different functions for the e-IDC—geotechnical investigations, engineering, construction, and monitoring, management and water balance—constituting the chain of elements responsible for success/failure of a dam.

Various hazards lurking within each element—for example: insufficient effort, mistakes, excessive audacity, etc.—leading to a probability of failure p_f for each element evaluated using a reliability model (Eqs. 4.1, 4.2) The chained elements can then be evaluated using, again, a reliability model (series, Eq. 4.1) in the case depicted in Fig. 4.3). The list of selected hazards should be discussed project by project.

Modes of failure previously identified in the literature [e.g. by United Nations Environmental Protection (UNEP); International Commission On Dams (ICOLD); World Information Service of Energy (WISE); Commission On Large Dams (USCOLD) and United States Environmental Protection Agency (USEPA)] due to the hazards selected in Fig. 4.3 are (Fig. 4.4):

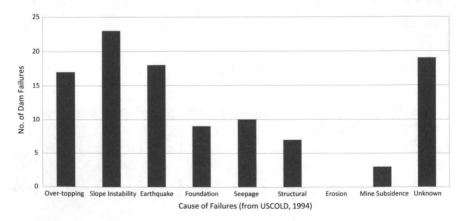

Fig. 4.4 Dam failures versus cause of failure. From USCOLD 1994

- Slope Instability;
- Earthquake and Mine Subsidence;
- Over-topping;
- Foundation;
- Seepage and Structural.

The Data

Data on tailings failure is scarce, sometimes tainted by biases and censorship, and spread through various entities and databases of variable reliability (for example, notice in Fig. 4.4 the very large number of "unknown" causes). In response to this we adopted a "quick and dirty" engineering approach to estimates, preferring to rapidly gain an understanding for the order of magnitude of the estimate rather than waiting to get very precise "true" numbers. We saluted the actuarial effort published in 2015 (Bowker and Chambers 2015) and were delighted to notice that our previous estimates were in good agreement with the more precise numbers, although we commented on some unfortunately "forced" linear regressions drafted by various authors and to the tendency to use variable time intervals to jump to conclusions.

In this section we ensure coherence with our earlier, now proven correct, "quick and dirty" engineering approach, but decided to also include uncertainties by using two different sets of causal lists, namely those resulting from ICOLD 1994 and those from a 1910–2009 compilation (Azam and Li 2010).

Failures Reported by ICOLD 1994

For the sake of this discussion it was necessary to re-interpret the data in the literature. Readily available records generally report number of failures versus cause of failure (Figs. 4.4, 4.5) and are fraught by many "unknown causes" or statements like "unusual weather" which allow plenty of room for conjecture. This discussion makes it necessary to attribute causality of the failures to the various phases

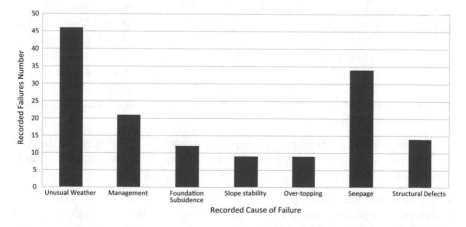

Fig. 4.5 Dam failures versus cause of failure from various sources (Azam and Li 2010)

Table 4.2 Recorded cause of failure, attributed causality scenario and related recorded failures number (from USCOLD 1994, Fig. 4.4)

Recorded cause of failure	Attributed causality scenario	Recorded failures number
Slope instability, earthquake and mine subsidence	Engineering error and omission, excessive audacity	$23 + 3 + 18 = 44$
Over-topping	Poor management (mostly in this example as a life time flaw rather than an initial one)	17
Foundation	Poor investigation	9
Seepage and structural	Poor construction	$10 + 7 = 17$
		Total $= 87$

of the e-IDC process, in order for the model to become amenable to analyses. Of course, should detailed data on causality of failures become available in the future, they could be readily included in the model and many assumptions made could be released/replaced.

Table 4.2 shows that out of a total of 106 recorded failures (Fig. 4.4), 87 are from known recorded causes (column 1), which were re-interpreted by us (column 2).

Failures World-Wide 1910–2009 Data
The data in Fig. 4.5 are a compilation (Azam and Li 2010) from UNEP, ICOLD, WISE, USCOLD, USEPA.

Table 4.3 shows the same re-interpretation described above regarding the data of ICOLD 1994 was performed for the 1910–2009 data. Out of 167 recorded failures, 145 are from known causes and 22 from unknown causes (column 1), which were re-interpreted (column 2).

The Relative Split of Attributed Causality Scenarios for Initial Flaws
Using the data of Tables 4.2 and 4.3 it is possible to define a relative (%) split of attributed causality scenarios stemming from project inception (Table 4.4). For ICOLD 1994 (Table 4.2) over-topping had to be removed (as it was attributed to management during lifetime), leaving us with 70 recorded failures with "known" causes. For world-wide 1910–2009 data (Table 4.3) we eliminate poor management for the same reason. Quake remains in the tally because if the dam fails under seismic loading, then the design should be considered as faulty from the beginning with respect to that loading.

We note a rather wide difference in the percentage split of causality, due to the poor quality of the database, requiring the study to proceed with both values to include uncertainties.

The Mitigations Models
In this study we consider two possible types of mitigations to be implemented during the e-IDC development: M1, independent peer review; and M2, inspections. These are described as follows:

Table 4.3 Recorded cause of failure, attributed causality scenario and related recorded failures number (Azam and Li 2010, Fig. 4.5)

Recorded cause of failure	Attributed causality scenario	Recorded failures number
Slope stability	Engineering error and omission, excessive audacity	9
50% unusual weather	Engineering error and omission, excessive audacity	23
	Engineering error and omission, excessive audacity	*Total of 2 lines = 32*
Management	Poor management (mostly considered in this example as a life time flaw rather than an initial one)	21
Over-topping	Poor management (mostly considered in this example as a life time flaw rather than an initial one)	9
50% unusual weather	Poor management (mostly considered in this example as a life time flaw rather than an initial one)	23
	Poor management	*Total of 3 lines = 53*
Foundation subsidence	Poor investigation	*Total of 1 line = 12*
Structural defects	Poor construction	14
Seepage	Poor construction	34
	Poor construction	*Total of 2 lines = 48*
		Total = 145

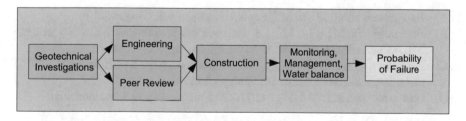

Fig. 4.6 In case (B) the base case is mitigated with M1 = peer review during engineering

- **M1**: Engineering performance can be enhanced with an independent peer review (including sensible RBDM procedures and risk assessment from project inception). Engineering and peer review become a parallel subsystem (Eq. 4.2, Fig. 4.6) possibly fraught by common cause failure (CCF).
- **M2**: Monitoring, Management, Water balance function can be enhanced with Inspections paired to sensible risk based decision-making procedures and risk assessment from project inception. Again CCF has nefarious potential on this additional parallel subsystem (Eq. 4.2, Fig. 4.7).

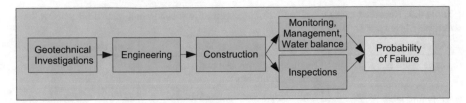

Fig. 4.7 In case (C) the base case is mitigated with M2 = inspections during service life

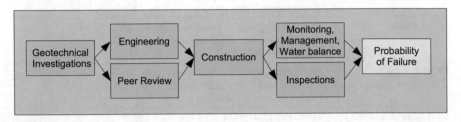

Fig. 4.8 In case (D) the base case is mitigated with M1 = peer review AND M2 = inspections

By adopting a very simplified approach to CCF it is possible to assume that insufficient rigour, complacency, conflict of interest, common excessive audacious approach in M1 and M2 could reduce the expected positive result of any mitigation to nil.

Four levels of mitigation (Fig. 4.3, i.e., the base case, Fig. 4.6, 4.7, 4.8) were studied:

(A) Base case with no M1 or M2, depicted in Sect. 4.2.3, Fig. 4.3.
(B) Base case with M1 (Fig. 4.6), i.e., independent peer review (including sensible RBDM procedures and risk assessment from project inception).
(C) Base case with M2 (Fig. 4.7), i.e., Inspections paired with sensible RBDM procedures and risk assessment from project inception.
(D) Base case with M1, M2 (Fig. 4.8) (descriptions as above) implemented.

By using ICOLD 1994 and World-wide 1910–2009 data (Tables 4.2 and 4.3) and the various mitigation variants of the model described above it is possible to evaluate the probability of failure of a dam under the considered hazard selection for a selected average life span of 30 years. In order to perform the calculations, one further step is necessary, as the probability of each hazard hitting a function has to be determined.

The first framing is easy in a professional environment: all those probabilities lie in the range 10^{-2}–10^{-4}. The higher value corresponds to a threshold where insurers generally shy away from insuring (thus any engineering/construction accident likely has a lower probability of occurrence), and 10^{-4} is a rate one order of magnitude above the upper bound of credibility (as engineering/construction accidents are unfortunately well within the credible realm). The second framing requires calibration of the model based on the data derived causalities (Tables 4.2, 4.3, and 4.4). Finally, the probabilities have to be annualised.

Table 4.4 Attributed causality scenario stemming from project inception

Attributed causality scenario	ICOLD		World-wide 1910–2009 data	
	Recorded failures	Failures in %	Recorded failures	Failures in %
Poor investigation	9	13	12	13
Engineering error and omission, excessive audacity	44	63	32	35
Poor construction	17	24	48	52
Total	70	100	92	100

Results
One Dam, Various Levels of Mitigation

Figure 4.9 depicts the results of the analyses for the four levels of mitigation (A), (B), (C) and (D) (Figs. 4.3, 4.6, 4.7, 4.8). In Fig. 4.9 the horizontal dotted/dashed lines "1979" and "2000" depict the framing estimates thresholds we published in (Oboni and Oboni 2013) for those decades. It is interesting to compare the results of the various mitigative levels (A)–(D) to those thresholds.

Base Cases (A) and (B): Base Case and Peer Review (including sensible RBDM procedures and risk assessment from project inception) have calculated probabilities of failure in the vicinity of 10^{-3}, i.e., the factual estimated rate we published in (Oboni and Oboni 2013) for the decade around 1979.

Case (C): Base case and inspections (over the life of the structure paired with sensible RBDM procedures and risk assessment from project inception) lead to a value near the mid-point of the values for the decades 1979 and 2000, bordering with

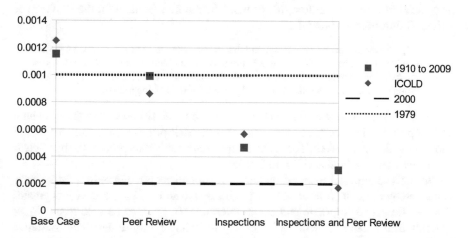

Fig. 4.9 Results of the annual probability of failures derived from the model for mitigation level A, B, C, D including attributed causalities (Tables 4.2, 4.3) derived from factual data (Figs. 4.4, 4.5)

the higher estimate from the most recent data (Oboni and Oboni 2016b; Bowker and Chambers 2015), for the Serious and Very Serious failures.

Finally, Case (D) with peer review and inspections reaches the lower bound of the interval, i.e., the value we published for the decade of the year 2000, and very similar to the lower estimate from the most recent data (Oboni and Oboni 2016b; Bowker and Chambers 2015) for the Very Serious failures.

Absent or botched mitigations M1 and/or M2 can of course increase the value of the probability of failure to historic high (decade around 1979) due to CCF.

To sum up, based on this example, the biggest reduction in the probability of failure of the e-IDC chain is obtained through thorough inspections with sensible RBDM procedures and risk assessment from project inception. Peer review has also a beneficial effect, of course, but smaller, probably because the most deviances and shortcuts intervene during the long term service life rather than during design. All together the implementation of both mitigation reduces the e-IDC chain probability of failure by almost one order of magnitude.

Prioritizing Risks in a Portfolio of Dams

Notwithstanding the assumptions made, which could be perfected in a real-life portfolio study, the model is capable of reconstructing first estimates (a priori) of the probability of failure in good agreement with the last one hundred years of tailing dams failure history by looking at data (records) that should still be available for many structures, possibly in corporate, governmental or regulators' offices archives. Thus, based on an examination of those records it is possible to determine, dam by dam, the first estimate of the probability of failure which, paired with the potential consequences each dam failure (to be determined using a multi-dimensional conse-quence analysis, see Chap. 12), will give the total risk and finally a dam portfolio (corporate, national, regional) quantitative risk prioritization. That quantitative risk prioritization would be the first step of what the Auditor General for the Province of British Columbia recommends:

> 1.10 Risk-based approach. We recommend that government develop a risk-based approach to compliance verification activities, where frequency of inspections are based on risks, such as industry's non-compliance record, industry's financial state, and industry's activities (e.g., expansion), as well as risks related to seasonal variations. (Bellringer 2016)

We have already demonstrated (Oboni and Oboni 2012) how to include societal and corporate tolerance (See Chap. 13) into the risk prioritization techniques and shown how decision makers' focus can be enhanced rather than obfuscated by unclear risk assessments (Oboni and Oboni 2016a).

Given the public and corporate capital investments required to reduce future risks generated by dams, it is paramount to first be able to address the situations of greater risk that lie above corporate and social tolerance. In order to be able to use more efficient prioritization it will be necessary to define multi-dimensional tolerance levels, an exercise that we have already performed from local to country-wide scale but cannot be discussed here due to limits of space.

4.3 Conclusions

Given the nature of tailings dams, their construction time and expected service life and closure, the effects of today's risk mitigation programs will only slowly become visible because the world portfolio will contain mitigated and unmitigated (legacy) dams. During that period the industry credibility and SLO will remain vulnerable (Oboni and Oboni 2014a, b; Oboni et al. 2013). It will be very difficult to evaluate progress as factors such as climate change, seismicity (again, not necessarily "Black Swans"), increase in population and environmental awareness (consequence side of the risk equation) will further complicate the situation. Thus public outcry and hostility toward the mining industry, fuelled by the diffusion of information and communication technology will likely increase unless transparent, rational, and defensible approaches to dam portfolio risk prioritization are swiftly implemented.

The model presented has been shown to "construct" the first estimate of the probability of failure of a dam which is consistent with factual historical world-data. As such it constitutes the first support to any prioritization effort on a portfolio of dams or a first attempt for a single dam. The causality of various factors entering in the e-IDC process and other elements in the dam's service life can then be individually discussed/negotiated among experts and stakeholders with a sensitivity analysis allowing for better communication and enhancing transparency.

It has been shown with a practical example where and how e-IDC process mitigative actions can be most beneficial, if properly implemented. The potential effects of CCF have been described. This methodical approach makes it possible to determine in an economical, orderly, efficient way the relative a priori probabilities of failure of dams, based on archival data, expressed in ranges to include uncertainties. It is possible to use this approach for one dam or for a portfolio of dams. Later on in this book (Chap. 14 and Part III) we will discuss more detailed approaches that require more data, but are still sustainable if portfolio prioritization is required with a "normal" level of data availability. Companies, governments, regulatory agencies dealing with large portfolios of dams need to be able to prioritize risks in order to develop credible and efficient risk reduction programs (Bellringer 2016).

References

Azam S, Li Q (2010) Tailings Dam Failures: A Review of the Last One Hundred Years, Waste Geo Technics

Bellringer C (2016) An Audit of Compliance and Enforcement of the Mining Sector, Auditor General for the Province of British Columbia

Bowker LN, Chambers DM (2015) The risk, public liability, & economics of Tailings Storage Facility failures, the Earthwork Action, 1–56

Caldwell J, Oboni F, Oboni C (2015) Tailings Facility Failures in 2014 and an Update on Failure Statistics, Tailings and Mine Waste 2015, Vancouver, Canada, October 25–28 2015. https://open.library.ubc.ca/media/download/pdf/59368/1.0320843/5

Canadian Dam Association (CDA) (2014) Application of dam safety guidelines to mining dams. Technical Bulletin

Chapman C, Ward S (2011) The Probability-impact grid - a tool that needs scrapping, in: How to manage Project Opportunity and Risk, 3rd ed., ch. 2, pp 49–51, Chichester, GB, John Wiley & Sons

Comar C (1987) Risk: A pragmatic de minimis approach, in: De Minimis Risk, ed C. Whipple, pp. xiii–xiv, New York, Plenum Press

Cox, LA Jr (2008) What's Wrong with Risk Matrices? Risk Analysis 28(2): 497–512.

Dezfuli H, Kelly D, Smith C, Vedros K, Galyean W (2009) Bayesian Inference for NASA Probabilistic Risk and Reliability Analysis, NASA/SP-2011-3421. https://ntrs.nasa.gov/archive/nasa/casi.ntrs.nasa.gov/20120001369.pdf

[FAA 2002] Federal Aviation Administration (2002) Aviation industry overview fiscal year 2001, FAA Office of Aviation Policy and Plans, Washington, DC

Hubbard D (2009) Worse than Useless. The most popular risk assessment method and Why it doesn't work, in: The Failure of Risk Management. Ch. 7, pp. 117–144, Hoboken, Wiley & Sons

Mahesh P (2014) Statistical Analysis of Common Cause Failure Data to Support Safety and Reliability Analysis of Nuclear Plant Systems, The Canadian Nuclear Safety Commission

[Mount Polley Report 2015] Independent Expert Engineering Investigation and Review Panel (2015) Report on Mount Polley Tailings Storage Facility Breach. https://www.mountpolleyreviewpanel.ca/final-report

[NASA 2007] NASA. NASA Systems Engineering Handbook SP-2007-6105. National Aeronautics and Space Administration. Washington, D.C. Chapter 6.4 (2007)

Oboni C, Oboni F (2012) Is it true that PIGs Fly when Evaluating Risks of Tailings Management Systems?, Mining 2012, Keystone CO

Oboni C, Oboni F (2013) Factual and Foreseeable Reliability of Tailings Dams and Nuclear Reactors—a Societal Acceptability Perspective, Tailings and Mine Waste 2013, Banff, AB, November 6 to 9, 2013

Oboni F, Oboni C (2014a) Ethics and transparent risk communication start with proper risk assessment methodologies, EGU General Assembly 2014, Vienna, May, 2014

Oboni C, Oboni F (2014b) Aspects of Risk Tolerability, Manageable vs. Unmanageable Risks in Relation to Critical Decisions, Perpetuity Projects, Public Opposition, Geohazards 6 (2014), Kingston (ON), Canada, June 15–18, 2014.

Oboni F, Oboni C (2016a) Military Grade Risk Application For projects' Defence, Resilience and Optimization, Risk and Resilience 2016, Vancouver Canada, November 13–16, 2016

Oboni F, Oboni C (2016b) A systemic look at tailings dams failure process, Tailings and Mine Waste 2016, Keystone CO, USA, October 2–5, 2016. https://www.riskope.com/wp-content/uploads/2016/10/Paper-17_A-systemic-look-at-TD-failure-TMW-2016_07-11_revised.pdf

[Oboni et al. 2013] Oboni F, Oboni C, Zabolotniuk S (2013) Can We Stop Misrepresenting Reality to the Public? CIM 2013, Toronto. https://www.riskope.com/wp-content/uploads/Can-We-Stop-Misrepresenting-Reality-to-the-Public.pdf

[Oboni et al. 2014] Oboni F, Oboni C, Caldwell J (2014) Risk assessment of the long-term performance of closed tailings, Tailings and Mine Waste 2014, Keystone CO, USA, October 5–8, 2014

[Oboni et al. 2016] Oboni F, Caldwell J, Oboni C (2016) Ten Rules For Preparing Sensible Risk Assessments, Risk and Resilience 2016, Vancouver Canada, November 13–16, 2016. https://www.riskope.com/wp-content/uploads/2016/11/Ten-Rules-For-Sensible-Risk-Assessments.pdf

Renshaw FM (1990) A Major Accident Prevention Program, Plant/Operations Progress 9(3): 194–197

Roche C, Thygesen K, Baker, E (eds.) (2017) Mine Tailings Storage: Safety Is No Accident. A UNEP Rapid Response Assessment. United Nations Environment Programme and GRID-Arendal, Nairobi and Arendal, ISBN: 978-82-7701-170-7. http://www.grida.no/publications/383

Stott JE, Britton P, Ring RW, Hark F, Hatfield GS (2010) Common Cause Failure Modeling: Aerospace versus Nuclear. In: 10th International Probabilistic Safety Assessment and Management Conference, Seattle, WA

Taguchi G (2014) Fault Tree Analysis of Slurry and Dewatered Tailings Management—A Frame Work, Master's thesis, University of British of Columbia

[TSM Task Force 2015] Report of the TSM Tailings Review Task Force (2015). https://mining.ca/documents/report-tsm-tailings-review-task-force/

Wilson AC, Crouch E (1982) Risk/Benefit Analysis, Cambridge MA, Ballinger Publishing Company

Chapter 5
What the Public Wants; Public Reactions

This chapter analyzes public reactions and "what the public wants", around the world, especially from the point of view of potential victims of failures.

In this chapter we attempt to define what a risk assessment should include using the results of a recent conference, Public Hearing in the NWT related to the environmental rehabilitation of the Giant Mine, as well as other information from the world.

5.1 Lessons Learned from Recent Conferences

The topics in this text are presented to provide a broader context for certain subjects discussed later on. The reader may or may not agree with what is stated but will, at least, better understand the basis of related statements or comments.

5.1.1 CIM 2015 Workshop

The State of Practice for Tailings Management Workshop at CIM 2015 Montreal discussed the Mount Polley Failure. This section summarizes part of the conversations and presentations which took place, as we interpret them. The workshop started with a Panel, Ministry, Legal, CIM and CAD positions' review.

As described earlier (Sect. 2.1), the Panel was given the mandate to define a path to zero failures, obviously a "political statement" more than a technical one, as it is well understood across all industries that "zero" is an unattainable objective. Thus, given the "unalterable wording" of the mandate the Panel had to carefully formulate

© Springer Nature Switzerland AG 2020
F. Oboni and C. Oboni, *Tailings Dam Management for the Twenty-First Century*,
https://doi.org/10.1007/978-3-030-19447-5_5

their conclusions. However, during the discussion several questions came up dealing with:

- The scant space given to management in the panel conclusions—due to time and scope limitation, following the replies given.
- How to prevent investigations from "missing" critical geotechnical/geological features? Are Factors of safety too aggressively selected? This will remain the responsibility of the proponents/owners following the replies given. CDA responded to the Mount Polley (http://www.cbc.ca/news/canada/british-columbia/mount-polley-mining-fears-1.4235913#_blank) failure by strengthening a number of statements and definitions (also on factor of safety, a point we have raised in a recent paper, when looking to the long term fate of dams (http://www.riskope.com/wp-content/uploads/2013/06/Risk-assessment-of-the-long-term-performance-of-closed-tailings.pdf#_blank). They recognized however that flexibility has to remain and that "one definition fits all" does not work.
- How to stop misrepresenting reality to the public (https://www.riskope.com/wp-content/uploads/Can-We-Stop-Misrepresenting-Reality-to-the-Public.pdf).
  The question was bearing on the wording of the mandate itself. "Path to zero failure" gives the public the illusion that the goal is reachable, whereas it is not. No industry in this world has ever reached such a goal. Very strictly regulated industries, such as commercial aviation, are carefully formulating finite goals in recognition of this. For example, in the United States the FAA aimed at reducing the commercial air carrier fatalities per 100 million persons on board by 24% over a 9-year period (2010–2018) reaching no more than 6.2 in 2018. NB: that is 6.2×10^{-8}, a value at least two orders of magnitude below the credibility threshold $(10^{-5}-10^{-6})$ many industries accept.

A reply stated that there is a hiatus between "risk management" and "safety" approaches. The reason is that safety management strives for "zero fatalities". In a private discussion afterwards we stressed the fact that the "zero fatalities" is certainly a honourable goal and that, while it works well as a slogan, it is a simplification of a complex reality. It keeps key personnel accountable and alert, but is certainly not a scientifically acceptable statement.

We again heard participants using the term "risk" as synonymous for a number of other technical words, and thus we again stressed the need for a proper glossary that specialists should use to avoid misunderstandings.

Finally, there was a specific request for clarity on defining tolerance to risk and enhancing understanding of complex consequences. It is obvious that this question cannot find a reply as long as common practice Probability Impact Graphs (PIGs) risk matrix and other fallacious methodologies are used in the mining (and other) industries, as pointed out in one of our 2012 papers (http://www.riskope.com/wp-content/uploads/Is-it-true-that-PIGs-fly-when-evaluating-risks-of-tailings-management-systems.pdf#_blank).

5.1.2 Managing Risks London Conference

Franco Oboni had the honour of chairing a session at the "Managing Risks Across the Mining and Oil and Gas Lifecycle" conference in London in July 2017. The session's title was "Managing our wastes for the long term" (https://www.riskope.com/2017/07/19/managing-risks-across-the-mining-and-oil-gas-lifecycle/). The conference was held at the Imperial College as part of The Geological Society's "Year of Risk" with the cooperation of the Institute of Risk Management. The conference's aim was to bring together mining and oil and gas professionals to discuss their respective and common risks, including how the industries manage their wastes for the long term.

We found it frustrating to see that, despite a unanimous agreement that changes and enhancements from common practices are badly needed, everyone seems to prefer status quo. Recurring phrases heard at the conference were: "we would like to…, but for this reason…, we can't" or "we would like to…, but we do not know how". In fact, both difficulties can be overcome by abandoning obsolete recipes, leaving behind common risk assessment practices, using scientific methods to evaluate holistic risks, and including social and environmental consequences, as will be discussed.

A case history was presented and the groups were asked to come up, as we will see later, with the preliminary steps of a risk assessment, i.e., performance criteria, system definition and hazard identification. The outcome was extremely interesting and clearly showed the asymmetry in the requirements of the two groups. It became apparent to all participants that reaching a consensus early on in the risk assessment procedure represents a viable step towards mitigating later chances of confrontation and dismissal of the results.

The common thread among the presentations was that systems have to be described and defined before sensible risk assessment and decision-making can be undertaken. Managing risks across the mining and oil and gas lifecycle requires, to start with, holistic approaches and a well-defined glossary. That came together with other points, summarized as follows:

- Simplistic solutions can have serious consequences, at both the modelling level and the mitigation level.
- There is major confusion even about basic definitions of risk and uncertainty, even within a single industry, or, even worse, across industries. It is time to see a single, robust glossary used world-wide.
- Both industries need approaches that are simple but not simplistic.
- We should not use words such as "never", "nil", and "zero", especially when talking about hazards and risks.
- Transparent credibility levels (e.g. 10^{-5} 10^{-6}) should be used.
- What is generally considered "long term" may be very short, especially in the face of events driven by climate change. And perpetuity may be an awfully long time.
- Censoring and biasing reality are lurking "mortal" sins.
- CSR and SLO dictate ethical behaviour and correct evaluation and presentation of risks.

If a risk assessment procedure bypasses all the above pitfalls and follows the standardized procedure that we will discuss in this book, then we consider it a rational risk assessment.

5.2   Public Hearing in the NWT: Environmental Rehabilitation of the Giant Mine

> There is a remarkable coherence in the societal requirements for better risk assessments, world-wide.

The text below was extracted from the MacKenzie Valley Review Board's "Report of Environmental Assessment and Reasons for Decision" (MVRB 2013: Appendix D, Specific Risk Assessment Requirements) and we reproduce it below with minor editorial changes.

It is expected a rational and societally acceptable risk assessment will include:

- The compilation of a proper glossary containing a description of all the terms used in the Project and its development, especially those that might have a common use which differs from the technical meaning (such as "risk", "crisis", "hazard") in compliance with ISO 31000.[1]
- The definition of the Project context in compliance with ISO 31000, including all the assumptions on the Project environment, chronology etc.
- A properly defined hazard and risk register.
- A clearly-defined system of macro and subsystems/elements and their functional and interdependent links describing for each one of them:

  - expected performances;
  - possible failure modes and quantification of the related ranges (to include uncertainties) of probabilities[2] evaluated as numbers in the range 0–1 (mathematical characterization) with a clear explanation of the assumptions underlying their determination;
  - associated magnitude of the hazards and related scenarios.

- An independent analysis of failure/success objectives.[3]

---

[1] In order to avoid any confusion, this text complies with the freely available glossary Riskope has produced. The glossary is now at his third edition (https://www.riskope.com/knowledge-centre/tool-box/glossary/).

[2] For the determination of the probabilities the assessors will use a selection of methods taken from ISO 31010 international code.

[3] For example, see NASA's Fault Tree Handbook with Aerospace Applications (NASA 2002). This should model success and failure with various pre-selected criteria.

- A holistic consequence function integrating all health and safety,[4] environmental, economic and financial direct and indirect effects.
- Applicable published correlations[5] and information.
- It is expected the risk assessment will use a unified metric showing consequence as a function of all health and safety, environmental, economic and financial direct and indirect effects/dimensions. This will be done in a manner that allows transparent comparison of holistic risks with the selected tolerance threshold.
- Consequences will be expressed as ranges, to include uncertainties. When evaluating the consequences, the risk assessment will:

  - Explicitly define risk acceptability/tolerance thresholds, in compliance with ISO 31000 international code (see Sect. 14.1). These will be determined in consultation with potentially affected communities, using a unified metric compatible with the one described above for consequences.
  - Risks[6] and tolerance or acceptability will be developed separately, in such a way not to influence or bias judgment of the assessors or evaluators. Risks will then be grouped into "tolerable" and "intolerable" classes. The risks in the intolerable group will be ranked as a function of their intolerable part. Mitigation efforts will be allotted proportionally to that ranking.

5.3 Sendai Framework

The so-called Global Platform was established for implementing the Sendai Framework for Disaster Risk Reduction (UNISDR 2015) adopted in Sendai, Japan, in 2015. In fact, the 2017 Global Platform conference (https://www.unisdr.org/conferences/2017/globalplatform/en) stated that there continues to be real risk (they used this wording) coming from political attitudes. For example:

- indicators will continue to measure disasters costs in dollars only;
- failure to take into full account the implications on the health, culture, environment, customs and ways of life of affected social groups;
- best technologies are used to identify the arrival of drought or the poisoning of water, earth and air, but without the powers necessary to mitigate them.

In fact we were very pleased to see that health, culture, environment, customs and ways of life of affected social groups are all considered to be valid failure criteria that should be taken into account in the multidimensional consequences of potential

---

[4]Including inference of casualties and pathologies deducted from health studies described in Measure 9.

[5]For example, using Holmes and Rae empirical correlation between "life changing units" and the likelihood to become ill due to external changes, stress, Societal Willingness to Pay (See Sect. 13.1.2), etc.

[6]Risks = (probability (range) * Consequence (range)) as an extension of Eq. (1.1).

accidents (http://www.riskope.com/2016/12/06/geohazards-caused-human-activity/
#_blank).

As a matter of fact, in (Oboni et al. 2013) we wrote that the public distrust towards
the mining industry originated in the fact that consequences are oftentimes poorly
defined. We also wrote that we should take into account "indirect/life-changing"
effects on population and other social aspects. These can be grasped by modern
techniques using simplified method and considering the wide uncertainties that surrounds the driving parameters. Among these:

- human;
- fish, fauna and top-soil/vegetation consequences;
- long term economic and development consequences;
- social impacts.

From the Sendai Framework we infer that it is likely that peoples and communities
will recover confidence in institutions if there is clear evidence of the willingness of
States to guarantee the right to life. This means that States would have to enhance
the work for effective regulation and protection of people and that industry will have
to adapt and respect international agreements and covenants.

We can note here the similarity between the Sendai Framework and the
Aashukan Declaration (https://aashukandotcom.files.wordpress.com/2017/04/the-
aashukan-declaration.pdf#_blank) of April 2017.

It seems again that preserving SLO and showing leadership in CSR require risk
assessments to become transparent, analyse the complexities of consequences and
allow transparent dialogue between stakeholders.

5.4 A Note on Communication and Transparency

5.4.1 Communication

As far as the public is concerned "Trust us" no longer works. As stated in the introduction, the social license to receive a permit to construct and operate (SLO) a tailings
facility depends nowadays on a company's willingness to engage in meaningful communication with the public with the objective of gaining their trust (https://www.
riskope.com/2017/06/28/stakeholders-not-satisfied-probability-assessments/). This
view is supported by the Australian Government, which states:

> A key challenge for mining companies is to earn the trust of the communities in which they
> operate and to gain the support and approval of stakeholders to carry out the business of
> mining. A 'social license to operate' can only be earned and preserved if mining projects are
> planned, implemented and operated by incorporating meaningful consultation with stake-
> holders, in particular with the host communities. The decision-making process, including
> where possible the technical design process, should involve relevant interest groups, from
> the initial stages of project conceptualization right through the mine's life and beyond. (AG
> TMH 2016)

Stakeholder consultation, information sharing and dialogue should occur throughout the TSF design, operation and closure phases, so viewpoints, concerns and expectations can be identified and considered. Regular, meaningful engagement between the company and affected communities is particularly important for developing trust and preventing conflict.

It should be noted that the term "consultation" is only one aspect of a meaningful communication program by a company. According to The International Association for Public Participation (IAP2), community engagement consists of a spectrum of approaches described as follows:

- inform (provide information);
- consult (obtain feedback);
- involve (act on what we hear);
- collaborate (public participates in decision-making process but company makes the final decision);
- empower (public decides).

The fourth level, "collaboration", closely parallels one of MAC's "leadership" level requirements as part of its "Towards Sustainable Mining" initiative. In its Aboriginal and Community Outreach Protocol (MAC Outreach 2015) one of the requirements at their leadership level is that formal mechanisms are in place to ensure that the public "...can effectively participate in issues and influence decisions that may interest or affect them".

Meaningful communication also requires that a company demonstrate its commitment by making its assessment protocols and results publicly available. In addition to helping to drive internal improvement, this practice will go a long way towards earning public trust by showing the comprehensive nature of the standards of practice being used and the efforts being made to ensure that they provide ongoing protection for the public and the environment.

To foster meaningful communication and engagement leading practice requires that the public is adequately informed of the nature of the risks relating to proposed and existing tailings facilities and can effectively participate, in a collaborative manner, in decisions that may interest or affect them. Leading practice also requires that a company make its assessment protocols and reports available to the public to foster transparency.

5.4.2 Transparency and NI43-101

To convince the government and the public beyond a reasonable doubt that the proposed site selection and deposition method provides an acceptable level of risk protection, a company must fully disclose the nature of the risks and demonstrate that its risk management strategies and its commitment to a strong governance framework will adequately address their concerns (Oboni and Oboni 2019).

This will require that the results of a dam breach and inundation study be disclosed and its risk mitigation measures be described. It will require that information be provided that supports the selection of the proposed alternative based on operating and closure requirements. Furthermore, as recommended in the British Columbia Guidance Document:

> Selection indicators for large projects should be conducted in consultation with local communities, First Nations, and stakeholders in order to maintain a transparent, defensible evaluation. (BC Guide 2016)

There are two main benefits of a meaningful communication process. The first is that by listening to and collaborating with the public regarding their concerns, a company will have a better appreciation of the risk mitigation measures (https://www.riskope.com/2015/08/27/integrating-mitigations-with-redundancies-in-an-industrial-process/) it should adopt. The second is that a company will gain the opportunity to demonstrate its commitment to high governance and risk management standards in a constructive manner and, if done right, set the basis for it to earn the SLO.

For new alternatives to be credible they must be supported by a high level of design, operating and closure expertise similar to that currently available for slurry deposition. They must also receive the highest level of corporate governance as the new technologies will present their own challenges and require greater attention to design assumptions and operating controls.

5.5 A Note on Ethics

5.5.1 General Ethics

In this section we are going to change the point of observation and ask ourselves what qualities should a risk assessment method possess to support ethical projects and endeavours? A wonderful book by Warwick Fox entitled *A Theory of General Ethics* (Fox 2006) can help us answering that question. The book proposes a "silo-breaking" theory of ethics. Through our respective professional roles, we are adamant in fostering silo-breaking attitudes in risk management. We are also strong believers in the need for geoethics (https://www.riskope.com/2017/08/23/virtual-reality-geoethics/#_blank), transparency in mining (NI-43-101) (https://www.riskope.com/2017/05/10/technical-session-towards-improving-environmental-social-disclosure-ni43-101/#_blank) and sensible risk assessments (https://www.riskope.com/2017/06/28/stakeholders-not-satisfied-probability-assessments/#_blank), i.e., to reach the goal for the twenty-first century. In this section we discuss how the General Ethics Theory can be linked to our practice and help the beneficial "silo-breaking" (https://www.riskope.com/2015/01/29/toward-the-disappearance-of-silos-effects/#_blank) endeavour.

Fox starts his discourse by stating that at the core of ethics lies the definition of the values we should live by. Philosophers refer to this as "normative ethics", the aim is to define the norms and standards that we should meet or attempt to meet in our conduct. Traditional thinking focused on human and their relationships with each other. That is an inter-human ethics, i.e., an anthropocentric view of ethics, disregarding, for example animal suffering. Ethics really started transcending inter-human ethics in the 1970s. There was the desire for a holistic approach of ethics to replace "silos" such as anthropocentric (humans), bio-centric (animals), eco-centric (environment) ethics.

The point is that the world, our "environment" consist of natural "self-organizing" systems as well as intentionally organized systems, built environments, resulting from the Anthropocene. Fox argues that General Ethics is a single integrated approach covering inter-human, natural environment and human constructed environment ethics.

The Theory of General Ethics, which Fox also calls "the theory of responsive cohesion", attempts to address three questions:

- What values should we live by?
- Why should we live by those values?
- How should we live by those values?

To answer but the first, Fox explains we need to define the "basic" or "fundamental" values in a single value statement. The most fundamental value there is—i.e., the foundational value—consists in a basic relational quality, or form of organization, which can be described as responsive cohesion. Responsive cohesion is one of three possible kinds of cohesion. It is the most important and valuable. The other two kinds are fixed cohesion and discohesion. Thus responsive cohesion is normative of relational qualities and tells us what relational qualities ought to be respected.

This summary will most likely leave you with appetite for more information. You can of course read the book, but before you do let us get back to our main subject with some examples.

The fixed cohesion characteristics are rigid, frozen, mechanical and prone to routine. Fixed cohesion is based on compliance, audits, codes and the application of fixed rules. Discohesion is chaotic, anarchic, "all over the place", "with no logic". Responsive cohesion "holds together" a system. Elements of the system respond to each other in deep, significant and meaningful ways. A responsive cohesion system is internally and contextually responsive. It is flexible, adaptive and creative. Using Fox's terms, its salient elements and features "keep company with each other", mixing predictability and surprise. The system is genuinely complex, organic and "alive".

Now that we have defined a responsive cohesion system and that we know that responsive cohesion is the foundational value, we also know that an ethical system has to "live" by selecting responsive cohesion maximization.

If we look at areas such as understanding and managing risks (maintaining an internally and contextually responsive system); designing rational risk mitigation based on Risk Informed Decisions (flexible solutions based on evolving uncertainties and hazards); updating on a regular basis system structure and hazards, risks (adapting mitigations); and Information transparency, SLO and CSR (integrating inter-human ethics with the other aspects), we see that it is indeed possible to develop procedures to abide to General Ethics. A scalable, drillable, updatable, convergent risk assessment methodology has the necessary qualities to support Ethics. Classic 4×4 or 5×5 risk matrix approaches (FMEAs, PIGs) do not have those qualities and contribute to create and maintain blind spots (https://www.riskope.com/2018/01/10/what-is-the-validity-of-your-risk-forecast/#_blank).

It has been reported (Radford 2009) that a group of Australian architects (Williamson, Radford and Bennets, University of Adelaide) have shown how they can apply the theory of responsive cohesion in the conception, design and construction of architectural systems. They apply the theory of responsive cohesion to buildings. We can envision the day where we will design large infrastructure projects, mining operations as responsive cohesive systems, thus fostering General Ethics.

The principle of cohesive response entails responding to ecological, social and built contexts in that order of priority. Then such a project will create sustainable value. It will require a significant reorganization of the design workflow. Projects will start with risk assessments and environmental impacts evaluations and then investigate and solve technical aspects.

5.5.2 Ethics and Tailings Risk Assessment

New tailings management approaches have to marry the requirements presented in the UNEP report mentioned previously, "Mine Tailings Storage: Safety Is No Accident" (Roche et al. 2017) while offering unparalleled support to independent risk assessors. The UNEP report identifies its requirement in distinct ways, by stating, for example: "Establish independent waste review boards to conduct and publish independent technical reviews prior to, during construction or modification, and throughout the lifespan of tailings storage facilities."

The report then adds:

- "Ensure any project assessment or expansion publishes all externalized costs, with an independent life-of-mine sustainability cost-benefit analysis." Including, of course the risks.
- "Require detailed and ongoing evaluations of potential failure modes, residual risks and perpetual management costs of tailings storage facilities."
- "Reduce risk of dam failure by providing independent expert oversight" done by independent risk assessor to maintain good and unbiased oversight. This will "ensure best practice in tailings management, monitoring and rehabilitation".

The independent risk assessor will ensure a drastic reduction of conflict of interest and the delivery of unbiased risk reports based on facts provided by modern monitoring techniques and transparent inclusion of uncertainties in the risk analyses. New tailings management approaches should deliver unbiased data interpreted using auditable rules, transparent risk registers. All requirements of UNEP will be met.

Following UNEP recommendations, independent risk assessor basing their assessment on auditable, repeatable selections of parameters has to become the new norm. These actions foster geoethics, will deliver a more defensible stance and value to the mining industry.

5.6 Conclusions

- The big picture of what people, the public and some international organization want is quite clear. "Trust us" does not work.
- Decisions made by any group that do not respect stringent technical requirements, that do not bring solid answers, and do not respect basic principles of ethics do not work anymore.
- Documents like the Aashukan declaration or the public record from the NWT Giant mine debate set the path for what the public requires now and in the decades to come.
- The price to pay for those who do not follow the path will be endless difficulties, fall of SLO and poor CSR performance.
- Fostering proper communication, with a clear language and a strictly defined glossary is a fundamental first step in the twenty-first-century tailings dams management.

References

[AG TMH 2016] Australian Government (2016) Tailings Management: Leading Practice Sustainable Development Program for the Mining Industry. https://www.industry.gov.au/sites/g/files/net3906/f/July%202018/document/pdf/tailings-management.pdf

[BC Guide 2016] British Columbia (2016) Guidance Document: Health, Safety and Reclamation Code for Mines in British Columbia. Version 1.0, July 2016. https://www2.gov.bc.ca/assets/gov/farming-natural-resources-and-industry/mineral-exploration-mining/documents/health-and-safety/part_10_guidance_doc_10_20july_2016.pdf

Fox WA (2006) Theory of General Ethics: Human Relationships, Nature, and the Built Environment, Cambridge, MA, MIT Press

[MAC Outreach 2015] Mining Association of Canada (2015) Towards Sustainable Mining Aboriginal and Community Outreach Protocol. MAC. http://mining.ca/towards-sustainable-mining/protocols-frameworks/aboriginal-and-community-outreach

[MVRB 2013] MacKenzie Valley Review Board (2013) Report of Environmental Assessment and Reasons for Decision Giant Mine Remediation Project http://reviewboard.ca/upload/project_document/EA0809-001_Giant_Report_of_Environmental_Assessment_June_20_2013.PDF

[NASA 2002] NASA (2002) Fault Tree Handbook with Aerospace Applications. https://elibrary.gsfc.nasa.gov/_assets/doclibBidder/tech_docs/25.%20NASA_Fault_Tree_Handbook_with_Aerospace_Applications%20-%20Copy.pdf

Oboni CH, Oboni F (2019) Reality-Anchored Risk Landscape Support CSR And SLO, MetSoc the 58th annual Conference of Metallurgists (COM 2019) hosting Copper 2019, Vancouver, Canada, August 18–21, 2019

[Oboni et al. 2013] Oboni F, Oboni C, Zabolotniuk S (2013) Can We Stop Misrepresenting Reality to the Public?, CIM 2013, Toronto. https://www.riskope.com/wp-content/uploads/Can-We-Stop-Misrepresenting-Reality-to-the-Public.pdf

Radford A (2009), Responsive cohesion as the foundational value in architecture, The Journal of Architecture, 14(4): 511–532 (2009) https://doi.org/10.1080/13602360903119553

Roche C, Thygesen K, Baker, E (eds.) (2017) Mine Tailings Storage: Safety Is No Accident. A UNEP Rapid Response Assessment. United Nations Environment Programme and GRID-Arendal, Nairobi and Arendal, ISBN: 978-82-7701-170-7. http://www.grida.no/publications/383

[UNISDR 2015] United Nations Office for Disaster Risk Reduction (2015) Sendai Framework for Disaster Risk Reduction 2015–2030. https://www.preventionweb.net/files/43291_sendaiframeworkfordrren.pdf

Chapter 6
Justifying the Need for New Approaches

This chapter draws the conclusions from the documents of the prior Chapters and shows why new approaches are indeed needed for Twenty-first century tailings dams management.

The topics in this chapter are presented to provide an understanding of the need for the identification and application of those advanced risk assessment and risk management methodologies required to support the strategic intent leading to "zero failures".

In particular, our purpose here is to describe and build on the insights provided by the reports of recent system failures described in earlier chapters, industry responses and other perspectives in order to identify those areas where the application of advanced risk-assessment methodologies can contribute to significant improvements in the performance of tailings management systems. The first step will be achieved by looking at the hazards that were unique to the moment of each accident and to identify those aspects of the systems that allow those causal factors to exist in the first place. From this analysis, supported by the perspectives provided by industry organizations and others, it will possible to identify the key areas where decisions during all stages of design, construction, operation and closure must be informed by a rigorous and disciplined approach utilizing advanced risk assessments methodologies.

6.1 Mount Polley Panel: Lesson Learned

The findings, observations, comments and recommendations provided by the Panel's report (Sect. 2.1) establish a solid basis for the identification of the type of decisions and approaches that must be developed for application on a broader scale in order to significantly advance the performance of tailings storage facilities.

© Springer Nature Switzerland AG 2020
F. Oboni and C. Oboni, *Tailings Dam Management for the Twenty-First Century*,
https://doi.org/10.1007/978-3-030-19447-5_6

6.1.1   Factor of Safety (FoS)

The first observation is that just because a given design has been created by a geotechnical firm using a stipulated FoS, and then adopted by a mining company and approved by a government, it is not in itself sufficient in itself to ensure that both the consequences and likelihood of failure meet the tests of materiality to a company and tolerance for the public.

With specific reference to the Mount Polley breach, it is noted that the FoS approach used for the first design did not lead to the recognition of the full nature of the foundation geology in the breach area. Furthermore the FoS approach was used to justify decisions to defer construction of the buttress and steepen the downstream slope from 2H:1V to 1.3H:1V. Both these alterations have been shown by the Panel to have had a significant effect on the robustness and resilience of the dam and, as such, significantly increased the likelihood of failure.

Another hazardous element that does not seem to have been adequately addressed by the FoS approach was related to the volume of impounded water, which had increased from 3.6 to 10.2 million m^3 in the three-year period prior to the breach. Less hazardous operating practices that would have reduced the volume of water would have, according to the Panel, possibly allowed for "timely intervention", thus reducing the likelihood of failure, or would have reduced the consequences of failure by "less than one third of what was actually lost". Both the expert panel and the Chief Inspector found that the Mount Polley tailings dam failed because the strength and location of a layer of clay underneath the dam was not taken into account in the design or in subsequent dam raises. The Chief Inspector summed up this situation by observing that other factors, including the slope of the perimeter embankment, inadequate water management, insufficient beaches and a sub-excavation at the outside toe of the dam, exacerbated the collapse of the dam and the ensuing environmental damage.

In its report, the Panel recognized limitations presented by the use of FoS as included in current Canadian Dam Association (CDA) Guidelines and the use of these guidelines as sole statutory requirements "… intended to be protective of public safety, environmental and cultural values …". The Panel also stated that tailings dam guidelines and criteria tailored to conditions in the Province of British Colombia would more effectively meet the needs of the province and added that this would "result in more prescriptive requirements for site investigation, failure mode recognition, selection of design properties, and specification of factors of safety."

6.1.2   Design Decisions and Approvals

The Panel rightly pointed out that key design and approval decisions must be based on a high level of detail that consider all technical, environmental, social and economic aspects of the project in sufficient detail to support an investment decision. In this regard the Panel emphasized the need for (1) comprehensive assessments of potential geotechnical problems; (2) a detailed evaluation of all potential failure modes; and (3) a management scheme for all residual risk. This highlights the need to base design decisions and approvals on a thorough identification assessment of risks within the context of the multidimensional aspects of the consequences (see Chap. 12) and the complexity of failure modes (Chaps. 9–11), as will be discussed further in this book.

6.2 Mount Polley Auditor General: Lessons Learned

The Auditor General's report does not elaborate on the general principles stated above but it does identify the need for a rigorous and disciplined risk-based approach by the regulator in keeping with the objective to protect the public from significant risks. If one starts with the premise that the public needs to be informed about the potential risks, it logically follows that the regulator need to be informed, permit applications need to be justified on the basis of risk tolerance and designs, and operating plans must explicitly identify and address residual risks. Such an approach would provide a solid basis for a strong compliance and enforcement program, which in itself would ensure a higher level of risk identification and management in the design of new mines and periodic permit amendments and would help focus ongoing compliance activities on the issues of importance.

Another contribution of the Auditor General's report relates to the degree of reliance that a government places on the information provided by engineers of record (EoRs) or qualified professionals retained by a company and the standards used to support their submissions. In particular, the report notes that an over reliance on qualified professionals was observed and that the acceptance of the industry design standard, as provided by the Canadian Dam Association's Dam Safety Guidelines (CDA 2007), was inadequate to guide both inspectors and industry. Both of these issues point to the need for government to have the ability to undertake or commission independent reviews, particularly those related to risk identification and management.

6.3 Samarco Panel: Lesson Learned

The first question that is raised by the Fundão Panel's report (Sect. 2.2) is related to the high level of detail and study that was required to analyse and understand what went wrong. That is, is an equivalent level of examination required for the initial design of a tailings dam and for subsequent modifications or changes to a design? The obvious answer to that question is that it would depend on the significance of the hazards that such designs or changes would be exposed to or generate. This answer can only be provided by the requirement for comprehensive change management procedures supported by rigorous risk assessment methodologies.

A second observation concerns the complexities related to the design and operation of the dam. The possibility of lack of adherence to procedures and quality control cannot be ignored and must be examined as part of a risk assessment process that takes a rigorous approach to hazard identification and an understanding of the uncertainties. Reports in the Brazilian press made reference to the use of FMEA in support of the design changes but, in whatever manner was used, it failed to identify the hazards created.

Given the complexities that have to be considered in the determination of the failure mode and its trigger(s), the Fundão Panel employed fault tree analysis to portray the many possible events and conditions as well as the various trigger mechanisms to guide its deliberations. The figure they display in their report provides an excellent example not only of the complex failure modes but also domino effects and long chains of events.

We can now focus on some legal and societal aspects starting with the common practice risk assessment and the decisions it stirred internally. The Samarco Fundão Dam failure mode and effects analysis (FMEA) (https://www.riskope.com/2015/01/15/failure-modes-and-effects-analysis-fmea-risk-methodology/) became the centre of an article in the *Gazeta Online* (2016) which described an exchange of messages by Samarco's internal communication system, reportedly between the company president at the time of the breakup of the Fundão Dam, Ricardo Vescovi, and various directors. In the exchange they discussed the use of FMEA. Another newspaper, the *Folha de São Paulo* (2016) reportedly had access to the transcript, made with judicial authorization and presented in the final report of the Federal Police about the tragedy of November 5, which left 18 dead and 1 missing. As a preamble we acknowledge that Mr. Vescovi's lawyer wrote in a note to the newspaper that

> the Federal Police investigation report is a provisional document based on unilateral understanding ... Ricardo Vescovi never received any notice or warning of possible impairment of safety Fundão Dam. He did not try to hide information of any sort. On the contrary, the information received about incidents, natural operation, indicated that the dam was within the safety standards. That was in accordance with statements by various specialists (our trans).

However, from the transcripts it becomes apparent that management of the mining company knew of the problems already in August 2014, more than a year before the disaster. Top management stated to Brazilian Federal Police that the instability problems were unknown to them. They also stated operation's managers dealt with them.

A discussion on the Fundão Dam reliability took place in 2011, following the same sources. Talking about the results of the FMEA of the Fundão Dam, Management asked whether "the likelihood of any problems happening had changed or just the severity (stiffness structure)?".[1]

We will pause here to note that, due to the arbitrary nature of FMEA (PIGs, risk matrix) cells limits, management would have the greatest difficulty to understand if

---

[1]Original Portuguese: *Vescovi indaga se "mudou a probabilidade (de acontecer algum problema) ou apenas a severidade (a rigidez a estrutura)?".*

the likelihood had really changed. We also note that the glossary was confusing as the severity does not have, at first sight, any link with the "stiffness" of a structure, unless a lengthy explanation would state that the "stiffness" is related to the velocity of development of a breach …

The transcript continues. "I think this point is the most important of all. It allows to show that things have not got worse. We are just being more critical in the assessment of severity".[2]

FMEAs are performed periodically as a way to monitor physical conditions of dams, but they do not necessarily hep management focusing on relevant issues (https://www.riskope.com/2016/02/24/80-20-rule-in-risk-management-practice-a-way-out-of-the-overwhelming-syndrome/#_blank). They could help defining operation strategy and tactical planning, but they lack the required finesse. They should also bring to light rational and unbiased prioritizations of risks, but, for a number of reasons explained elsewhere (https://www.riskope.com/wp-content/uploads/Can-We-Stop-Misrepresenting-Reality-to-the-Public.pdf), they generally do not.

At Samarco, management used them as a persuasion tool:

> It is worth bringing up in the text something that corroborates a low probability of an event. An FMEA for example, beyond the opinion of ITBR.[3]

The ITBR was the internal committee formed by Samarco employees and also external experts. ITBR's task was to evaluate the mining structures, with meetings every four months.

The courts will define who was legally at fault and liable in the case of the Samarco catastrophe. It is not to us to make any judgment or suggest any fault. That in particular to show respect to the victims, the families and all those involved in the accident.

6.4   Oroville IFT Lesson Learned

The IFT (Sect. 3.1) is commended on the scope and depth of their work. In particular, the IFT's detailed examination of the non-physical or, human, organizational, and industry factors that led to the failure of the spillway systems adds significantly to the understanding as to how risks are created and what has to be done to overcome the challenges or barriers related to the identification of hazards and their effective management.

---

[2]The original response in Portuguese: *Acho esse ponto o mais relevante de todos, pois é o meio de mostrarmos que as coisas não pioraram, apenas estamos sendo mais críticos na avaliação de severidade* (July 27, 2011, at 23h 58 in response to a question from the mine).

[3]The original text in Portuguese: *Vale a pena abordarmos no texto algo que corrobore com uma baixa probabilidade de um evento. Como o FMEA por exemplo, além da própria opinião do ITBR.*

6.4.1   Human Factors

The use of a "human factors" framework to guide the IFT's investigation of the Oroville incident was very useful by helping to identify many of the factors that have to be addressed or overcome in the establishment of a culture devoted to the highest standards of risk identification and management. The approach adopted by the IFT is described in detail in Appendix J of the IFT Report. Selected aspects of Appendix J are summarized below.

Based on observations by the IFT of past failures, it is stated that "… failures—in the sense of human intentions not being fulfilled—are fundamentally due to human factors." To protect against human failures, it is then observed that barriers, such as factors of safety, are established which must be overcome for failure to occur. From this it follows that those human factors that may contribute to failure must be identified and addressed.

The primary drivers of failure are described as pressure from non-safety goals, human fallibility and limitations and complexity. Complexity is described as:

> …resulting from multiple system components having interactions which may involve non-linearities, feedback loops, network effects, etc. Such interactions can result in large effects from small causes, including "tipping points" when thresholds are reached, and they make complex systems difficult to model, predict, and control… Complexity generally exacerbates the effects of human fallibility and limitations.

It is then stated that, while the drivers of failure lead to various types of human error, this:

> …is not a sufficient endpoint for a forensic investigation. Instead, human errors should be treated as prompts to investigate further and delve deeper, to understand the situational and contextual factors which contributed to those human errors…That is the approach the IFT endeavoured to take for this investigation.

With regard to risk management, the IFT states that the underlying primary drivers of failure that often lead to inadequate risk management are primarily due to: (1) Ignorance—not being sufficiently aware of risks; (2) Complacency—being sufficiently aware of risks but being overly risk tolerant; and/or (3) Overconfidence—being sufficiently aware of risks, over estimating ability to deal with them. While not stated as such by the IFT, these factors are basically barriers to good risk management which must be overcome.

The application of human factor methodology to the Oroville incident required the consideration of a wide scope of factors and how they could have an impact on judgments, decisions, actions, inactions, influencing situational factors, and interactions. In that the ITF's finding with respect to human factors are presented in detail in the main report, Appendix J presents an overview of some of the higher—level human factors that contributed to the spillway incident. The observations pertaining to the subject of this book are as follows:

> The inadequacies in dam safety risk management that contributed to the failure primarily involved ignorance about the existence of the risks associated with the spillways, which was

mainly due to insufficient expertise regarding spillway failure modes and mischaracterization of the geology. Confirmation bias related to perpetuating misunderstanding of the geologic conditions at the spillways likely contributed to this ignorance of risks. There was also a degree of complacency regarding tolerating risks, and some overconfidence in ability to manage risks.

Appendix J concludes by stating:

Overall, in terms of human factors, the safety "demands" which contributed to failure were significant, while the systemic "capacity" to meet those demands and maintain dam safety was substantially lacking in many areas. Half a century after design and construction of the spillways, this systemic imbalance in human factors had set the stage physically for the spillway incident to occur in February 2017.

The application of the human factors framework by the IFT has illustrated the need for its use not only in dam failure reviews but also for periodic integrity reviews and in support of a wide range of decisions, judgments, management system requirements and risk assessments. How advanced risk assessment methodology can provide the discipline and rigor to identify and address the human barriers to safety will be a major focus of this book.

6.4.2 ITF Lessons on Risk Identification and Management

The lessons summarized above regarding failure mode analysis, comprehensive facility reviews and physical inspections all emphasize the importance of risk identification and management. The identified weaknesses of the Potential Failure Mode Analysis (PFMA) approach are real and specific suggestions for improvement to the PFMA process were presented in Appendix F3 of the ITF report. Many valid suggestions were made, but it is not sufficient just to do a better job of PFMA's. A more rigorous and disciplined approach is needed. With regard to the suggestion that more comprehensive reviews are required, the added depth and detail that would be provided will establish a solid base for the effectiveness of whatever risk assessment methodology that may be applied.

With respect to physical inspections, regulatory compliance and owner's dam safety programs, a risk-based approach to support and guide these activities would definitely add value in terms of developing and effective surveillance and monitoring program and providing the basis for corporate and government oversight of the critical issues.

6.4.3 What Else Can the IFT Report Tell Us?

The IFT Report offers wealth of insights, observations, findings and lessons. If they could be compressed into one page, the key points might be as follows:

1. Failures—in the sense of human intentions not being fulfilled—are fundamentally due to human factors;
2. Failures are caused by a complex interaction of relatively common physical, human, organizational, and industry factors;
3. Numerous human, organizational, and industry factors lead to the physical factors not being adequately recognized and properly addressed;
4. "The dominant risks to be managed in dam safety derive not from unique events but from adverse combination of more usual events. One might think of these as unusual combinations of usual conditions … These include the systems interactions among management policy, procedural factors, instrumentation …, operational and maintenance practices, design flaws, construction compromises, deterioration, outside disturbances, and many other things that are often overlooked at the time of design. The result is that many system failures do not fit within a traditional engineering-analytic framework and a new systems framework is needed."
5. The new framework should:

   • Provide a thorough understanding of the wide scope of human factors that may detract from the degree of risk protection with particular attention to their impact on activities such as judgments, decisions, actions, inactions, influencing situational factors, and interactions;
   • Require a detailed understanding of the systems to be managed based periodic comprehensive reviews as the basis for reliable and quantified input to a risk assessment process. This would require more in-depth reviews than usually mandated and would include:
     – Current standards and states of practice;
     – A review of all available data;
     – Critical and independent thinking;
     – Completed by people with appropriate technical expertise, experience, and qualifications to cover all aspects of design, construction, maintenance, repair, and failure modes of the assets under consideration.
   • Utilize expanded risk assessment approaches that would address:
     – Difficulties in properly characterizing risks for large or complex systems, including accounting for human and operational aspects in failures;
     – The tendency to oversimplify complex failure modes involving multiple interactions of system components by defining failure modes as a linear chain of events;
     – The need to consider how broader organizational factors, such as culture and decision-making authority and practices, can contribute to failure.

6.4.4 A Risk-Based Perspective

The cause of the Oroville Dam spillway incident was a long-term systemic failure of the California Department of Water Resources (DWR), regulatory, and general industry practices to recognize and address inherent spillway design and construction weaknesses. In addition, poor bedrock quality, and deteriorated service spillway chute conditions caused the incident.

Indeed, this was a case of normalization of deviance over service life during which no one evaluated how the multi-hazard risk landscape of the system had evolved. Quantitative multi-hazard convergent, scalable and drillable risk assessments are paramount. Monitoring, near-misses and inspections should feed their updates.

Thus the incident cannot reasonably be "blamed" mainly on any one individual, group, or organization. As a matter of fact, failures does not happen overnight and the "common practice" siloed management structure is certainly to blame, rather than any individual, group or organization. Abolishing the silo culture is paramount.

There was no single root cause of the Oroville Dam spillway incident, nor was there a simple chain of events.

Interdependencies and common cause failures (https://www.riskope.com/2015/04/16/how-system-definition-and-interdependencies-allow-transparent-and-scalable-risk-assessments/#_blank) need to be accounted for in the risk register (i.e., complex failure modes, domino effects, long chains of events). The inclusion of complex failures modes would deliver to management a better perception of reality.

Responding to the damage to the service spillway chute necessitated difficult risk trade-offs in times of crisis and risk trade-offs may be extremely difficult when on top of actual uncertainties one finds a layer of "soft uncertainties" due to the lack of robust and uniform glossary, poor risk assessments, misleading conclusions. That becomes especially important in infrastructures with a long-term service life. Fostering clarity and transparency ensuring the use of robust and uniform glossary, training people to recognize hazards from risks, understanding well-made risk assessment, which are not using risk matrix-based approaches will go in the right direction.

At the end of the day decisions made without a full understanding of relative uncertainties and consequences, allowed the reservoir level to rise above the emergency spillway weir for the first time in the project's history, leading to severe and rapid erosion downstream of the weir and, ultimately, the evacuation order. Furthermore decisions made under stress, in fear and under poor information (https://www.riskope.com/2018/01/03/acting-intuitively-planning-360-view-upward-downward-tactical-strategic-risks/#_blank) are indeed oftentimes ill-guided and tend to look to the short term while forgetting the longer term. Risk assessment should describe the multi-hazard risk landscape, and deliver understanding of uncertainties and multi-dimensional consequences.

Following the unexpected chute slab failure and erosion, and subsequent closure of the service spillway gates to examine the damage, delicate and difficult risk trade-offs, involving myriad considerations, were necessary over the next few days in order to manage the incident. Risk-informed decision making (RIDM) and risk-based decision making (RBDM) makes it possible specifically to manage those considerations. RIDM and RBDM are preparatory processes, not crisis-time reactions. Indeed, they should rely on adaptive hazards and risk registers that can easily update.

More frequent physical inspections are not always sufficient to identify risks and manage safety. The inclusion of the quantification of mitigation efficiency in the risk landscape is paramount. That means performing residual risk assessments. The goal is to define a rational risk reduction roadmap.

Periodic comprehensive reviews of original design and construction and subsequent performance are imperative. These reviews should be based on complete records and need to be more in-depth than periodic general reviews (need for risk repository/documents databases), such as the FERC-mandated five-year reviews at the time of the incident.

As soon as conditions change (https://www.riskope.com/2017/08/08/10-commandments-for-resilient-design/#_blank) the risk landscape should reflect those changes. Qualified individuals must give attention to appurtenant structures associated with dams, such as spillways, outlet works, power plants, etc. This attention should be commensurate with the risks that the facilities pose to the public, the environment. Efforts should include risks associated with high consequence events which may not result in uncontrolled release of reservoirs.

Compliance with regulatory requirements is not sufficient to manage risk and meet dam owners' legal and ethical responsibilities. Some of these general lessons are self-evident. Others noted them before the IFT's investigation of this incident. Also, to maintain SLO and a good CSR, the mere following regulatory compliance criteria (https://www.riskope.com/2011/02/23/why-legal-negligence-test-is-not-a-critical-test-for-an-operation/#_blank) is inadequate. Fostering continuous communication and consultation including explanations of necessary risk trade-off is paramount.

6.5 Common Practice Balances and Checks

6.5.1 Audits, Assessments and Reviews

A survey of corporate sustainability reports, corporate websites and government requirements has shown very little commonality regarding the use of such terms. The use of the term audit is generally clear although some audits rely on the use of judgment to a qualified extent while other instances an activity described as an audit is more of a review. The terms assessments and reviews, and sometimes evaluations, are used interchangeably and sometimes together. Some companies refer to formal reviews or periodic reviews, which sounds good but gives no evidence of the substance of the activity suggested.

Audits are typically described as the independent, formal, systematic and documented examination of an organization's or facility's performance with explicit, agreed, prescribed criteria. To be effective audits need detailed protocols that provide specific questions to which factual answers can be provided as proof of conformance with practice requirements.

Assessments and reviews differ from audits primarily to the extent that judgment, based on relevant levels of experience and professional qualifications, including quantitative analyses of some kind, is used to evaluate the effectiveness of designs or practices in achieving desired outcomes. For the purposes of this present discussion a review is defined as a formal examination of something with the objective of verifying attainment of a required performance level and, if not so, identifying the need for improvement. In this context, assessment is defined as the process of making a judgment about something based on analyses. Assessment is primarily a tool to judge or quantify progress or lack thereof against an objective although possibilities for improvement can also be a valuable outcome. In this last case the assessment is used as a tool for risk informed decision making.

While the terminology used to describe a company's assurance activities needs better clarity, the effectiveness of any one activity depends on the establishment of a clear understanding of what is expected, what the main issues are and what qualifications are needed from the audit, review or assessment team. Finally the methodology used has to be proven effective and reliable in generating the required support.

The judgment and experience of a team must be matched with the objectives of the assurance activity. Judgment and experience levels will increase in inverse proportion to the availability of detailed protocols. Judgment and experience levels will be especially important when examining critical control procedures and practices related to high consequence risks. The quality of any given verification activity will always depend on the experience and judgment of the verifier and the quality of the verification protocol and applied methodologies.

6.5.2 Assurance Activities

Mining companies vary considerably in their approach to assurance activities. Larger companies have started to internalize some of their requirements for assurance through the establishment of corporate internal audit functions that are described as being independent. Such companies may also internalize technical and operating expertise for the development of corporate standards, to provide support for individual operations and to provide support for and be involved in internal assurance programs.

Smaller companies, which do not have the resources to establish a corporate internal audit function or to provide technical and operating guidance, have to rely on external providers to meet their policy requirements. For example, assurances related to an ISO 14001 (2015) management system are available from professional auditors and assurances related to technical design can be provided by geotechnical engineers. MAC members benefit from the assurance protocols and providers as part of their Towards Sustainable Mining (MAC TSM 2017) program.

Technical dam reviews on a periodic basis are generally required by governments and are also essential parts of a comprehensive corporate assurance program. One feature of the Legislated Dam Safety Reviews guideline published in British Columbia (APEGBC 2014) is its recognition that the level of assurance should depend on a number of site specific circumstances such as consequence rating, dam type and use. It also requires an assessment of "the operations, maintenance and surveillance practices at the *dam* including the assessment of the overall *dam* safety management system and identification of any non-conformances;" without providing any guidance relating the conduct this part of the dam review. This is just one example of the need to examine all assurance activities test for overlap and gaps as well as to the suitability of the supporting protocols and guidance's.

The main challenge for all companies in the establishment of a comprehensive and effective assurance program is to ensure a high level of technical and operating expertise is available, internally or externally, to assist in the development of the scope, terms of reference and the selection of suitably qualified professionals for each assurance activity. The second challenge is to ensure that adequate assurance protocols and guidance's are available to meet the objectives of the assurance activity. The third challenge is to ensure that their assurance activities also assess the quality of site-specific operating, monitoring, surveillance, maintenance and reporting procedures as described in the operating manual and audit their implementation and conformance in practice.

6.5.3   Monitoring

When and if a permit is approved it must be granted with the requirement that critical design operating parameters and risk mitigating strategies are strictly adhered to. One example is the requirement, as stated in the B.C. Guide (2016), to include measurable monitoring parameters that are identified and required to be maintained within predetermined limits for a tailings storage facility. This subject was also addressed in a report by the Auditor General of B.C. (BC AG 2016) that stated that permits should be written with enforceable language.

Companies should be required to provide notification of any proposed changes to the permitted deposition method, dam design or operating conditions for government review and approval. Such notifications and supporting material should be accompanied by the results of a quantitative risk assessment that clearly identifies any consequence or likelihood changes. The risk assessment should be carried out by a third party and should be drillable in such a way that risk comparisons are possible. If there is any doubt as to the acceptability of the revised risk profile, governments should require a public review as part of its approval process. Governments, at any time, should also establish the right to compel companies to provide an independent opinion on any proposed changes based on terms of reference approved by government, including appointing a third party independent risk review.

6.6 Synthesis of the Forensic Analyses, Public Opinion, Institutional Reports

In this section we build a "table" of specifications for twenty-first-century tailings dams management. The left column (Table 6.1) bears the "management" specs distilled from the previous chapters and sections, the right column the "risk evaluation" specs we have defined over the years. As a matter of fact, the right column is strongly inspired by a 2013 document entitled "Twenty rules for good

Table 6.1 Twenty rules for good risk assessment

Management specs	Risk evaluation specs
Start early with a comprehensive management system. Do not wait to see the first errors or near misses to start!	Start early. Often the best competitive advantage is brought by developing risk assessments already at pre-feasibility stage
There is the need for a rigorous and disciplined risk-based approach by the regulator in keeping with the objective to protect the public from significant risks	Hazards are anything that can go wrong You have to understand the context of the study and what constitutes the system you have to assess Time spent on defining the system is very well spent Time spent making sure the logic of your hazard register is correct is very well spent
	The largest and costlier mistakes are generally made when defining the system. Shortcuts and logical confusion (which are very common) are very expensive!
Regulators need to be informed, permit applications need to be justified on the basis of risk tolerance and designs and operating plans must explicitly identify and address residual risks	In modern society, he who hides risks dies, sooner or later If you want to stay out of jail, never use 0 or 1 for probabilities
	Zero risk does not exist Certainty does not exist Always explicitly deal with uncertainty. A range is far better and more credible than a single number
Factor of Safety is not an indication of soundness of design, resilience of the structures, reliability or risk exposure	The same hazard can provoke widely different risk because consequences can vary in time and location, even if the FoS is constant
FoS can even mislead if applied to mitigation decision making You need something else and that something else is: comprehensive assessments of potential geotechnical problems, a detailed evaluation of all potential failure modes, and a management scheme for all residual risk	Evaluate mitigations' risks, not only their reliability. Mitigations reduce probability of occurrence of scenarios, but they do not change consequences. Alterations of the system alter consequences

(continued)

Table 6.1 (continued)

Management specs	Risk evaluation specs
There is the need for governments to have the ability to undertake or commission independent reviews, particularly those related to risk identification and management	People involved in a process, be it a project, a company, a venture are not the best people to build a risk assessment, because they are biased. Third party intervention is absolutely necessary
Communication has to be fostered thorough a project and its risk assessment. Communication allows building trust among the parties	Don't ever be afraid to ask questions, like a child to people that know the system. Generally they have become "blind" to their risks
Consequences of hazard occurrences have to be studied thoroughly: it is not enough to define a few categories and then select the worse as often suggested in PIGs applications	Past can never be assumed to equal the future. At best it can be used as a point estimate among others. Consequences are always multidimensional and very with location, time, etc.
Primary drivers of failure that often lead to inadequate risk management are primarily due to: (1) Ignorance—not being sufficiently aware of risks	Probability Impact Graphs (PIGs), which constitute "common practices" in many industries, don't fly, because they are misleading, lend to biases and censoring and do not give a proper roadmap for future development and risk mitigations
(2) Complacency—being sufficiently aware of risks but being overly risk tolerant and/or optimistic	Unless you understand what is manageable versus Unmanageable, the future is going to hurt
(3) Overconfidence—being sufficiently aware of risks, over estimating ability to deal with them with arrogance	Manageable risks are the one that can be mitigated to become tolerable. Unmanageable risk cannot be brought to be tolerable unless the system is altered Tolerance has to be defined in order to allow proper decision making Tolerance definition requires transparent communication with stakeholders
As project evolve through their life there is a requirement for comprehensive change management procedures supported by rigorous risk assessment methodologies	Risk assessment updates have to be simple and information should be preserved and reused in the cyclical needed updates
The same as above goes for operations, monitoring and management, repairs	As above
Human factors that may contribute to failure must be identified and addressed as they are "hazards" like any other	Results cannot be delivered in a binary way: "this system is safe", or "risks are under control" do not work anymore
Complexities like inter-dependencies and their boosting effects have to be studied and evaluated	Risk assessment has to include inter-dependencies, common cause failures (CCF) and cover 360-view of the hazard and resulting risk landscape. Risk assessment has to be convergent (all hazards are looked at, no siloed information)

(continued)

Table 6.1 (continued)

Management specs	Risk evaluation specs
Physical inspections, regulatory compliance and owner's dam safety programs would benefit from a risk-based approach to support and guide these activities. They would definitely add value in terms of developing and effective surveillance and monitoring program and providing the basis for corporate and government oversight of the critical issues	Risk assessments, allow the design of sensible business continuity plans and emergency response plans. These need to be tested in order to ensure the system works as planned under duress. Always start testing and drilling with a table-top exercise and then keep increasing the difficulty until you reach the "surprise" test level
The dominant risks to be managed in dam safety derive not from unique events but from adverse combination of more usual events	No risk assessment should ever be performed (even of administrative processes) without covering natural hazards (including of course, climate), man-made hazards, cyber and technical hazards
Provide a thorough understanding of the wide scope of human factors that may detract from the degree of risk protection with particular attention to their impact on activities such as judgments, decisions, actions, inactions, influencing situational factors, and interactions	Talking about resiliency increase without first performing a comprehensive convergent risk assessment is like shooting in the dark. Lots of money will be wasted
Require a detailed understanding of the systems to be managed based on periodic comprehensive reviews as the basis for reliable and quantified input to a risk assessment process. This would require more in-depth reviews than usually mandated and would include: Current standards and states of practice; A review of all available data; Critical and independent thinking; Completed by people with appropriate technical expertise, experience, and qualifications to cover all aspects of design, construction, maintenance, repair, and failure modes of the assets under consideration	Avoid incestuous risk assessments, use third party competent and not "project-interested" expertise. Gather wide spectrum data beyond the traditional monitoring data the owner may provide
Utilize expanded risk assessment approaches that would address: Difficulties in properly characterizing risks for large or complex systems, including accounting for human and operational aspects in failures; The tendency to oversimplify complex failure modes involving multiple interactions of system components by defining failure modes as a linear chain of events; The need to consider how broader organizational factors, such as culture and decision-making authority and practices, can contribute to failure	These are the subjects treated in PART II and PART III of this present publication

risk assessments (https://www.riskope.com/wp-content/uploads/2013/06/Riskope-20-rules-for-good-Risk-Assessments.pdf)". Unless a management system follows these specs point by point and brings quantitative answers to risk prioritization the necessary leap forward in tailings dams management will not happen.

As a side note we add that what has been specified applies to any TSF project, including dry paste, co-mingled and any other technology that may be developed in the future. As we write we have already reviewed new projects including some of these technologies that "take more risks" as the materials are deemed superior to standard deposition methods ones. This is a never-ending loop which requires the greatest attention.

References

[APEGBC 2014] APEGBC 2014. Association of Professional Engineers and Geoscientists of British Columbia (2014) Professional Practice Guidelines – Legislated. Dam Safety Reviews in BC V2.0. https://www.apeg.bc.ca/For-Members/Professional-Practice/Professional-Practice-Guidelines

[BC AG 2016] Auditor General of British Columbia (2016) An Audit of Compliance and Enforcement of the Mining Sector http://www.bcauditor.com/pubs/2016/audit-compliance-and-enforcement-mining-sector

[BC Guide 2016] British Columbia (2016) Guidance Document: Health, Safety and Reclamation Code for Mines in British Columbia. Version 1.0, July 2016. https://www2.gov.bc.ca/assets/gov/farming-natural-resources-and-industry/mineral-exploration-mining/documents/health-and-safety/part_10_guidance_doc_10_20july_2016.pdf

[CDA 2007] Canadian Dam Association (2007) Dam Safety Guidelines

Folha de São Paulo (2016) Samarco cogitou remover vila de Mariana antes de tragédia em MG https://www1.folha.uol.com.br/cotidiano/2016/06/1784196-samarco-cogitou-remover-vila-de-mariana-antes-de-tragedia-em-mg.shtml

Gazeta Online (2016) Relatório da PF revela que ex-presidente da Samarco sabia de falhas em represa. https://www.gazetaonline.com.br/noticias/brasil/2016/06/relatorio-da-pf-revela-que-ex-presidente-da-samarco-sabia-de-falhas-em-represa-1013952121.html

[ISO 14001:2015] International Organization for Standardization (2015) ISO 14001:2015 Environmental Management Standard, International Standards Organization https://www.iso.org/obp/ui/#iso:std:iso:14001:ed-3:v1:en

[MAC TSM 2017] Mining Association of Canada (2017) Towards Sustainable Mining 101: A Primer http://mining.ca/towards-sustainable-mining

Part II
How to Build a Twenty-First-Century Comprehensive Risk Management Program

Part II, How to Build a Twenty-First-Century Comprehensive Risk Management Program, starts with a basic list of "DON'Ts." After that the tone turns positive and the text plan follows the logic of a real-life risk-informed decision-making approach. Indeed, each chapter corresponds to a necessary step to develop a modern risk assessment which can then lead to risk-informed decision making "from cradle to grave" of a tailings facility or any other mining system.

Chapter 7
Let's Start with Some Serious "Do Nots"!

7.1 Do Not Declare a Dam "Safe"

This chapter explains what not to do under the form of a list of DO NOTs.

As already shown earlier for companies, engineers, and regulators, to portray, describe, or assess a tailings dam in terms of its being safe is misleading and incomplete, as the case histories in Chap. 2 have already shown. Furthermore, this can be unethical and go against fostering SLO and CSR.

One of the main problems to be addressed is the complacency engendered by the use of the terms "safe" or "safety" when referring to dam designs and operating procedures. It may an excellent aspirational goal but, as discussed in Chap. 2, and long-term history (Chap. 4) shows that tailings dams are not necessarily as safe as some parties may claim. All dams are exposed to hazards posed by seismic and meteorological events and all are exposed to design and operational hazards that may be created by shortcomings, singly or in combination, with regard to the development and application of geotechnical knowledge, the level of corporate commitment and the strength of regulatory oversight.

A paradigm shift in the approach is necessary to alter the course of the events. A urban legend attributes to Albert Einstein the following: "We cannot solve our problems with the same thinking we used when we created them." Even if the quote was not from Einstein, we think it is wise enough to remember.

"Mine Tailings Storage: Safety Is No Accident" (Roche et al. 2017) follows the same lines of thought: A Rapid Response Assessment also identifies a common practice that has to stop. The developer or design-engineers self-risk assessment has to stop as it is fraught by conflict of interest.

© Springer Nature Switzerland AG 2020
F. Oboni and C. Oboni, *Tailings Dam Management for the Twenty-First Century*,
https://doi.org/10.1007/978-3-030-19447-5_7

While having a safe dam may be a noble objective, having a design or operating management plan declared safe is misleading to the public and to those who have to make decisions related to the integrity of individual tailings management facilities. We will now discuss several points related to the "safety" discussion.

7.2 Do Not Accept Incremental Answers

The design, construction, operation and closure of any tailings dam is carried out within a complex system requiring high levels of expertise, commitment and diligence. A mining company, which must accept ultimate responsibility, will generally retain professional consultants to assist them in meeting their responsibilities and may rely on the regulatory system to add rigor through government permitting and compliance responsibilities.

Whatever the situation, a complex relationship exists between mining companies, regulators and consultants with regard to ensuring that the highest standards of tailings dam design and management are identified and implemented. Many individual elements have to come together within a system that examines the interrelated nature of the activities that each must perform in meeting their respective responsibilities. For the purposes of this book, this integrated relationship will be called the Tailings Responsibility Framework (TRF).

Recent dam failures discussed in Part I (Chaps. 2 and 3) have led to the identification of incremental improvements in the design, management and regulation of tailings dams that should be adopted on a global basis to reduce the likelihood and consequences of future catastrophic failures. However, the mining industry must move beyond the incremental answers provided by the narrow "accident investigation" approach and use advanced risk assessment methodologies to identify the gaps, hidden flaws or hazards that have been created within existing TRFs that could contribute to future catastrophic failures.

7.3 Do Not Call Unforeseeable What Indeed Is Foreseeable

The immediate response to a dam failure is generally to call for an investigation to determine the cause of the failure. The cause is then established by expert consensus and generally described as a geotechnical/hydrogeological/hydraulic/structural failure of some parts of the dam system. It may then be traced back to a hazard that created a fatal flaw in the design, an operating practice that caused the integrity of the design to be compromised and/or the failure of government to enforce the conditions of its operating permit. A dam failure may be triggered by a particular natural or man-made event or action but the root cause of failure is not what happened at the moment. The root cause is what decisions were made that let that hazardous situation exist in the first place. In the case of a particular hazard or a combination of hazards, the fact that they were allowed to exist within the TRF needs to be examined from a number of perspectives as described below. Oftentimes the first reactions are to call any failure a "unforeseeable event", "unheard of", "impossible to predict because of the uncertainties, complexities", etc.

> There a consensus driven definition of what represents an "incredible" event.

At this point it becomes necessary to go back to basic definitions and to state what an "incredible" failure may be (Oboni and Oboni 2018). It is an important addition to the discussion when the concepts of negligence, often used in the aftermath of a disaster and come into play (https://www.riskope.com/2010/03/17/force-majeure-clauses-in-contracts-should-be-optimized-to-reduce-costs-and-litigation-potential/). As stated earlier, using a general consensus from various horizons of the hazmat industry we will consider that the threshold of credibility lies at around 10^{-5}–10^{-6} and that any event with probabilities of that order or lower can be considered "incredible", or an "Act of God".

7.4 Do Not Forget Uncertainties and Latencies

In this section we will discuss tailings risks from the point of view of the metaphors for public perception of risks developed by German researchers Renn and Klinke (2004), developed a series of metaphors to describe public perception of risks and establish a classification for some categories of tailings dams. More details are available in the specialized literature and we have written on this subject (https://www.riskope.com/2014/10/02/metaphoric-description-of-risks-and-their-perception-are-they-useful-or-just-wordplays/).

Although quite theoretical and apparently abstract, the classification of Renn and Klinke has the merit of explicitly using the level of uncertainties on probability and/or consequences for the definition of risks as well as the concept of latency.

> Here is a summary of Renn and Klinke's metaphors describing public perception of risks:
>
> **Sword of Damocles** has a very high disaster potential, with the probability oftentimes extremely low. Extremely well designed, built and managed tailings dams with high consequences belong to this category.
>
> In the risk class **Cyclops** the probability of occurrence is largely uncertain whereas the maximum damage can be estimated.

Pythias includes risks associated with the possibility of sudden non-linear behaviour, such as the risk of self-reinforcing global warming or of the instability of the West Antarctic ice sheet, with far more disastrous consequences than those of gradual climate change. Static liquefaction failures would belong to this category.

Pandora's Box is characterized by both uncertainty in the probability of occurrence and extent of damage (only presumptions) coupled with high persistency.

Cassandra is characterized by a relatively lengthy delay between the triggering event and the occurrence of damage. If the time interval were shorter, the regulatory authorities would certainly intervene because the risks would be clearly located in the intolerable area. Many tailings dams are in this category.

Medusa refers to the potential for public mobilization. This risk class is only of interest if there is a particularly large gap between lay risk perceptions and expert risk analysis findings. Poorly made risk assessments and poorly disciplined approach may transform tailings dams risks into this category.

Sometimes risk assessors experience difficulties (if they do not, then they are doing something wrong) when people balk at wide range of events scenarios, trying to bias the studies toward what they believe are "credible" events in terms of probability p and consequences C (losses). The censoring they propose is generally double:

- reducing averages and
- narrowing ranges to say −10 to +15%.

That censoring is of course completely arbitrarily-selected and based on "experience". That experience is unfortunately often short term and localized, thus useless or misleading.

For example, while assessing risk on mountainous mines' (tailings) pipelines and roads we consider a maximum scenario which includes full loss of the pipe and the road along it. Of course we look at local topography and other parameters before introducing such an event at a given location (Oboni and Oboni 2018).

That maximum scenario generally encounters fierce hostility from the client or the professionals in charge of operations. We have seen that reaction occurring both in North and South America. They consider that scenario as excessive, actually "incredible" based on their "experience and gut feeling", and they bring plenty of reasons for that "incredibility".

Yet these accidents occur and generate risks. They come loaded with potential fatalities and image/reputational consequences.

So, those accidents are not only occurring, but they occur in many countries around the world with a rate that oftentimes places the generated risks in the societally intolerable realm (see Chap. 13). As a result, those accidents are unfortunately not "incredible" and show that "experience and gut feeling" are not good advisors when assessing risks.

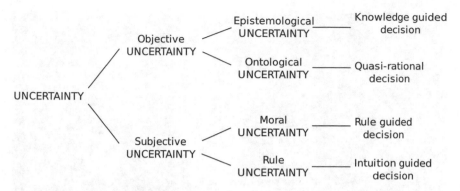

Fig. 7.1 A taxonomy of uncertainty

The concept of latency introduced by Renn and Klinke is a very important one with tailings dams, where errors and omissions, decisions made at project inception may generate failures decades into the structure service life.

Continuing the discussion on uncertainties, lately we found an interesting taxonomy of uncertainty in view of decision making (Fig. 7.1).

To make things simple, let's summarize the taxonomy this way:

The top half (Objective Uncertainty) of Fig. 7.1 is where risk analysis is.

Epistemology: is concerned with the nature of knowledge itself, its possibility, scope, and general basis. More broadly: How do we go about evaluating risks? or How do we separate true risks from "white noise"? or How can we be confident when we have analyzed the true risks? Rational quantitative risk assessments are epistemological applications as they are a systematic way by which we can determine when something is good or bad for a business and make knowledge-based decisions.

Ontology: is concerned with identifying, in the most general terms, the kinds of things that actually exist. When we ask deep questions about what is the hazardous environment of a facility? Or is there a risk? We are asking inherently ontological questions. Hazard identification is a ontological action, and PIGs risk matrix (https://www.riskope.com/2016/12/28/answering-risk-questions-cannot-answered-common-practice-approaches/) would also fall in this category, despite their well-known limitations, delivering quasi-rational decisions.

The bottom half (Subjective Uncertainty) (https://plato.stanford.edu/entries/formal-belief/) of Fig. 7.1 is where gut feelings and pre-conceived behavior lie.

Moral: here belong all the approaches based on pre-cooked lists and forms, audits, compliance actions. People will deliver Rule-based decisions, without understanding the existence and truth of the risk landscape surrounding the decision.

Rule: this is where gut feelings, your rules, i.e., pre-conceived behaviors, are kings. This is where, if you are lucky people will remember you as a "genius", but if you are wrong, people will forget about you. You will join the scores of audacious intuition-guided decision makers, or the "stupid guys" of Fig. 7.2.

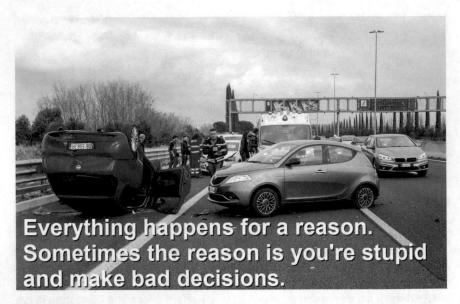

Fig. 7.2 "Everything happens for a reason. Sometimes the reason is you're stupid and make bad decisions." Marion G. Harmon, Ronin Games

We can draw three conclusions from this discussion:

Knowledge-guided decisions may not be flashy and prestigious, but are defensible and will increase your long term success record.

Intuition-based decision will give you an adrenaline fix and open two options: either glory (very seldom) or oblivion.

The above two points are clear as applied to a traffic accident (Fig. 7.2). So why do board rooms and high-stake decision makers often forget them?

In its own ironic way the quote in Fig. 7.2 reminds people about "The law of attraction (aka The secret)", a law stated by the so called New Though philosophy, and how we can influence our world, one way or another, with our decisions and, more importantly with the way we make them or what motivates them.

7.5   Do Not Jump to Risks: Hazards Come First!

A hazard is defined as a condition with the potential to cause undesirable consequences. Thus a hazard can be an event or a person or a group of persons, a behaviour, etc. with a certain likelihood of occurrence and potential consequences on the system. Natural hazards such as earthquakes and meteorological events are important but they may also be created due to a wide range of actions and decisions throughout the mining cycle related to technical, management and regulatory factors.

Technical hazards may be related to inadequacies in the geotechnical body of knowledge, unresolved uncertainties in the design process, lack of an appropriate level of engineering experience and judgment and unprofessional conduct. Management related hazards can stem from weak corporate commitment to the oversight of material issues, inadequate management systems, cost and production pressures, lack of adherence to key operating parameters and poor management. Regulatory hazards can be introduced as a result of design approvals and amendments being given without an adequate focus on risk tolerance, the lack of strict permit operating and construction conditions and the absence of an enforcement culture.

Human factors can also result in hazards not being recognized or addressed adequately (see Chap. 9). As stated earlier (see Sect. 6.4.1), the underlying drivers of failure that may lead to inadequate risk management are primarily due to: (1) Ignorance—not being sufficiently aware of risks; (2) Complacency—being sufficiently aware of risks but being overly risk tolerant; and/or (3) Overconfidence—being sufficiently aware of risks, over estimating ability to deal with them.

If the focus is not, at all times, on the identification and reduction of hazardous situations root causes, they may go unrecognised, uncertainties may not be understood, and resulting risks may not be properly managed. Here is an example: if, when asked to approve the development of a new mine, a company's board of directors is told that the design has been declared safe by their professional geotechnical consultants and it has been approved by the appropriate regulator, the corporate board members have little reason to probe further. If, however, the board members are told that:

(1) tailings management has been identified as a material/corporate risk, and the tailings system represents an intolerable but manageable risk exposure, with the intolerable risk being, for example, the fourth largest corporate risk;

(2) the consequence classification of the proposed dam location and deposition method evaluated in terms of lives, direct costs, environmental damages, share capital loss, reputation is the third within the corporate portfolio; and

(3) that a risk-based approach to mitigation has identified a road-map reducing risks to a corporate and societal tolerable level,

then a significant set of dynamics are put into play.

Back in 1955 the psychologists Joseph Luft and Harrington Ingham developed a tool called the Johari window (Luft and Ingham 1955). It splits the self-perception and public perception space in four areas. The Johari window became world-renowned when the then US Secretary of Defence Donald Rumsfeld quoted parts of it. You probably remember the known unknowns and the unknowns you do not know you don't know in an "intelligence" talk. There are interesting applications of the Johari window in risk management (https://www.riskope.com/2017/06/21/johari-window-application-risk-management/).

Risk assessments should start by recognizing that the absence of evidence is not evidence of absence: if something has "never" occurred it does not mean it does not exist.

The Arena is **"Known knowns"**. These are "facts" and likely the realm of service life/operational "incidents", stuff that we all know occurs time to time. These can also be uncertainties, when their probability is "one" and the only parameter that can vary is the magnitude.

Blind spots is where we find **"Known Unknowns"**. These are unexpected evolutions of known events: they should be considered as visible emerging risks.

Facade is where we find corporate **"Unknown Knowns"**, i.e., issues management prefers not to see, not to know. Companies that keep playing rosy scenarios and censoring their risks have a strong Facade.

Finally the true **Unknown**, i.e., the **"Unknown Unknowns"**. These are issues we truly do not know we do not know…and only swift adaptive management can solve these….i.e., working with a sustainable, highly resilient system.

Appropriate Risk Assessment and Management quite obviously expand the corporate **Arena** at the expenses of **Blind-spot** and **Unknown**.

These dynamics will result in a more disciplined and rigorous approach that will, if taken seriously, work towards significant improvements in the rational management of hazardous situations, hence risks related to the design and operation of tailings storage facilities. Uncertainties will be identified and reduced, remaining hazards will be addressed and critical control procedures will developed with the objective of mitigations of risks that may have been created by bad management practices.

Maintaining the status quo while merely changing the words to claim a goal of "zero failures", talking about robust, resilient and responsible mining practices will not do.

7.5.1 Dynamics Within the Tailings Responsibility Framework

Another way of looking at how hazards can go unrecognised or fail to be properly addressed is to examine the dynamics within the TRF and how each major participant may influence or respond to pressures within their respective organizations. As indicated earlier (see Sect. 1.2), the term regulatory capture has been used by the Auditor General of British Columbia to describe the situation where the regulator, created to act in the public interest related to the design and operation of tailings storage facilities, may, in certain situations, serve instead the interests of a company or the industry (BC AG 2016). Using the same perspective for mining companies, economic capture could be described as the situation where short or long term economic factors are given precedence over risk management, design parameters, sustainability commitments and permit obligations. For the geotechnical community, client capture could be used to describe situations where an engineer may be influenced by client pressure in the performance in their work.

7.5.2   Inadequate Standards

It must be recognized that effective documented governance, design and operating practices have been developed by highly committed companies that have achieved a high degree of risk awareness, which sometimes goes beyond what is generally called "best practices". It must also be recognised that some companies may not have a high degree of commitment; these require strong industry leadership, strong regulatory oversight and a high standard of professional ethics from the geotechnical profession.

As stated earlier, the quality designation for an industry or professional practice rests solely in the eye of the beholder, which is most likely to be the person or organization that has developed it. Common terms are "good practice", "best practice" and, less often, "leading practice". Even the term "emerging practice" has been used recently to describe a practice that has been employed by some committed companies for over 20 years.

The concern with most designations of this kind stems from how they were developed. As indicated earlier, at the lowest level, the designation is a self-identified by an individual, company or an organization, followed, going higher, by a committee of experts, usually under the umbrella of a mining association, government or professional body. Because of varying degrees of commitment at both the committee and approval level, this inevitably leads to a less than aggressive pushing of the boundaries. The next level of practice development occurs when other stakeholders are involved, as it has been shown that the resulting practices lead to the inclusion of more demanding and detailed expectations, such as more specific guidances and performance thresholds. The term "best practice" may be applicable to such situations.

Statements regarding leading practice often refer to instances where a practice exhibits a higher degree of commitment as compared to other companies, other associations or other political jurisdictions. Such definitions are essentially backward-looking. While undoubtedly some aspects may represent a high degree of commitment for specific practices, many still fall short of what constitutes real leadership. Leading practice should be forward looking and defined as a practice that goes well beyond the norm, providing the highest degree of commitment and/or a significant improvement in risk reduction.

While an understanding as to how industry practices are developed and agreed upon is important, the real test is how they are defined. Prescriptions can range from a list of things to think about (checklists, for example) to detailed descriptions as to what is required. Without a detailed description of what each practice element requires, little guidance is provided for both the practitioner and those that need to judge the level of commitment being applied. Again, as discussed in Chap. 6, experience has shown that stakeholder involvement has helped identify the need for more detail.

The bottom line is that existing self-identified good and best practices have failed to identify the hazards and related risks that existed prior to recent catastrophic dam failures and more progress is required. Furthermore, the more general the description of a practice, the easier it is for the less committed to claim they have adopted it. From another perspective, most industry standards, practices and guidelines can lead to complacency and overconfidence without the application of a rigorous and disciplined approach to risk assessment.

7.5.3   Closing Notes on Hazards

Factors of safety, industry standards, best practices, etc. are tools commonly used for the design and operation of tailings dams. They can offer a perceived high degree of protection against failure through the provision of high technical standards, fully-developed management systems and effective government oversight. Such contributions to risk reduction could be considered as strong barriers to failure or protective factors. However, over many decades they have been shown to be inadequate to meet the challenge and should not be accepted as the final answer. Their use can be compromised by the technical, management, regulatory and human hazards noted above, which may or may not exceed the ability of the protective factors to prevent failure.

The challenge is to move industry thinking beyond the complacency and over-confidence that may be generated by claims that existing standards and regulations have been applied. The mining industry must move beyond existing standards on a case-by-case basis to ensure that all hazards have been identified, adequate protective risk mitigation features have in fact been applied, and that a rigorous and disciplined approach has been applied to the elimination of any negative human factors that may exist.

Thus the proof is in the pudding. Unless risk assessor, geotechnical designers and mining companies seriously change their approaches to risk, things will continue to go on as they are today without notable improvements.

However, all believe their tailings designs and facilities are safe. In reality all dams are not created equal and some have failed. As stated in (Freeze 2000):

> The owners and operators of large engineering facilities want the public to hear about the great benefits to be bestowed upon it by their facility, not about how likely it is to fall down, or what the probability is that it will pollute the environment.

Note that, in the quote above, risk is implicitly mentioned, as the author spoke about p_f as "likely to fall down" and of a significantly censored C "probability it will pollute the environment" reduced to one dimension. Indeed C generally has a very complex metric including public safety, H&S, business interruption, environmental damages, image and legal damages, and crisis potential as already discussed as well.

The ultimate goal should be that a future mining project will not approved by a company's board of directors or by a government unless the proponent can demonstrate to itself, the government and the public, beyond reasonable doubt, that the proposed TSF can be managed in a manner that meets each party's definition of acceptable risk (see Chap. 13 for a discussion of risk tolerance).

For a company, its strategic vision should be to gain the confidence of the government and the public for its tailings management plans though the demonstration of a transparent, unbiased risk assessment, commitment to strong policies and practices that are capable of earning their trust and meeting their definition of acceptable risk.

Furthermore placing "safety first", as often stated, is a vague objective as long as it is not quantified. Hazardous industries and nuclear power plants clearly define their safety objectives. Tailings do not.

> Objectives have to be clearly stated and defined. They have to operable, considering the real life portfolio of dams.

This is particularly true when looking at the world-wide portfolio of dams at highly variable stages of lifecycle development. To be on the path of "zero-failure" we have to consider possible "congenital problems". These include, for example, insufficient depth of geotechnical investigations, insufficient factors of safety, etc. These problems will continue to affect the world-wide portfolio for decades unless significant mitigations are implemented. Only rational risk assessment and the clear definition of a safety/success criteria make it possible to prioritise those mitigations through RIDM and finally increase societal safety. Mitigative funds must be allotted rationally and sensibly, and, above all, sustainably, in order to correct the mid- and long-term situation.

> It is preposterous to find correlations with variables—for example, the price of copper—as some authors have claimed and use those to formulate predictions. That is unless it is assumed that the selected variable, for example the price of copper, generates the "straw that breaks the camel's back" situation. However, in that case the reasoning should go toward the root causes and not toward the last element of the chain.

Additionally, we note there is no definition of what a residual risk assessment should entail, as pointed out in a case history in Sect. 5.2 (MVRB 2013: Appendix D, June, 2013) and perpetual cost of waste storage facilities cannot be evaluated using the classic (common practice) NPV because of its very well-known drawbacks. Here too, solutions exist (see Sect. 15.10).

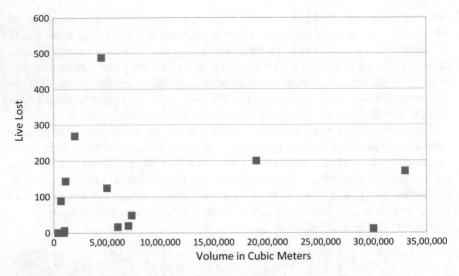

Fig. 7.3 Volume in m³ versus live lost in tailings accident from 1962 to 2014

As an example of "clouded misinformation" on hazards and risks, let's look at an intuitive assumption heard at conferences around the world, which states that the bigger the dam the bigger the consequences. It happens that statement is often proven wrong because of the multitude of parameters involved and the multidimensional aspect of failure. For example, as we can see in Fig. 7.3 the tailings volume doesn't correlate at all with the number of casualties (Caldwell et al. 2015); in other words, failure magnitude cannot be used to gut-estimate consequences, without proper and detailed analysis.

All communications, research, design and operating controls must be clearly focused on the likelihood and complete consequences of failure and the risk management strategies that are being or should be used to protect its employees, the public and the environment. The word "safety" will not be used in this book other than when referring to the safety of employees, the public or the environment at large.

7.6   Do Not Consider Consequences of Failures as One-Dimensional

No accident has one-dimensional consequences of failure which can be described with "one magic" number.

As already stated, the consequences of a dam failure are a function of multiple dimension such as environmental, harm to people, business interruption, legal, etc.

Any attempt to rate the consequences as a multiple "OR" clause (e.g., consider the worst among potential casualties, OR environmental, OR ...) is to be rejected as simplistic and prone to biasing the "safety" perception. Let's start with an example drawn from recent history: the cumulative damages effects of the Samarco dam disaster (already partially discussed earlier in Sect. 2.2).

We have read a great number of articles related to the true cost of the failure of Samarco dam. The dam failed in November 2015, killing at least 19 people, leaving 700 people homeless, and polluting hundreds of miles of the Rio Doce river, all the way to the Atlantic Ocean. However, we had to wait for the BHP annual report to see some details of the cumulative damages effects of the Samarco dam disaster.

In the document BHP Billiton detailed what the true potential costs might be and the list was as follows:

- 2.3B\$ in civil public actions filed by state prosecutors in Minas Gerais
- 3.1B\$ in damages filed by public defenders in Minas Gerais
- 620M\$ in damages filed by state prosecutors in Espirito Santo
- 6.2B\$ in a public civil claim
- 43B\$ for reparations, compensation, and "moral damages" in relation to the Samarco dam failure.

Oftentimes we see Failure Mode and Effects Analyses (FMEAs (http://www.riskope.com/2015/01/15/failure-modes-and-effects-analysis-fmea-risk-methodology/)) performed where the cost of consequences associated to each risks scenario are described with a single value deriving from a "OR" clause. As stated above, it is indeed common practice methodology (certainly not a recommended practice) to split the consequences in categories—for example reputational, environmental, cost and number of lives lost—then ask the risk assessor to select the worst and use that "single-value" for the determination of the risk. Over the years we have shown to many of our clients why this approach is bogus. The Samarco dam failure gives a perfect example:

- Cost of consequences associated to each risks scenario have a single value;
- Costs are additive once an hazard hits or a scenario occurs. Consequences are an additive function of at least the following costs:
  - direct and indirect,
  - health and safety,
  - environmental, image and reputation,
  - legal, etc.

As we can see in the Samarco list, the legal, the reparations (moral damages), etc. costs have now indeed been added. In the oil and gas arena, BP learned the lesson in the hard way with their 62USD billion additive estimate for the Gulf disaster. This is the reason why, when we perform any risk assessment, even a very simple one (http://www.riskope.com/2016/09/21/risk-assessment-comparison-system-including-5-macro-elements/), we use the consequences category listed above in an additive way!

Once it is understood that consequences are multidimensional it becomes self-evident that qualifying safety with a single number—be it a Factor of Safety (FoS) or a probability of failure p_f—is severely insufficient.

So, let's now start looking at the problem from the side of consequences Henry Brehaut used the term "materiality" in his keynote lecture at TMW 2017, and we will adopt it here.

- From a corporate internal perspective materiality is primarily defined in financial terms.
- For a government, materiality is largely defined in non-economic terms considering possible loss of life, environmental damage and economic loss (see Sect. 5.3).
- For the public, materiality is primarily defined in terms of personal impact with their personal safety being paramount.

The elements that must be considered for the definition of the overall metric of losses are, at least and not in any particular order: (a) direct and indirect; (b) health and safety; (c) environmental; (d) image and reputation; (e) legal, etc. (Oboni and Oboni 2016).

The potential economic losses to a company for even a partial failure of a tailings dam include loss of profits and costs related to dam reconstruction, environmental rehabilitation, lawsuits and government fines. Economic factors alone will usually dictate that tailings dams be considered a material risk for a corporation. In addition, a company must also consider potential impacts such as the loss of human life, environmental damage and public economic loss in its materiality ranking. When all potential impacts are considered, including a company's loss of public credibility, it is hard to visualize a company not considering the design and operation of any one of their tailing dams not to be a material risk issue.

7.7 Do Not Forget to Define Performance, Success and Failure Criteria

In addition to committing to a comprehensive risk assessment program, the mining industry should commit to ensuring that:

- Suitably qualified and experienced experts are involved in TSF risk identification and analysis, as well as in the development and review of effectiveness of the associated controls.
- Performance criteria are established for risk controls and their associated monitoring, internal reporting and verification activities (ICMM Position 2016: 5).

In our experience, very often mining companies skip a first, very important step: defining the failure criteria to be considered in the risk assessment. Indeed, without failure (or malfunction) there is no possible risk definition. Thus, from the very beginning this is paramount in order to avoid confusion, conflict of interest, biases, censoring from this very point on. Censoring and biases occur when, for example, experts convene to look only at "credible failures", without defining what "credible" means; they neglect to properly define the system (see Chap. 8) and use ill-conditioned binning exercises (FMEA, PIGs) with arbitrarily indexes and related values. Any "safe dam statement" based on a blatantly censored and biased scenario is obviously devoid of any meaning and unethical.

Part of this effort also goes in using a well-defined and clear glossary (https://www. riskope.com/knowledge-centre/tool-box/glossary/). A clear example comes from a relatively recent event in the Solomon Islands (https://www.riskope.com/2016/04/20/solomon-islands-gold-mine-contaminated-water-spill-disaster/). In the case of Solomon Islands' Gold Ridge dam, arsenic, cyanide and reportedly selenium and mercury are present in the tailings. Excess water accumulated behind the dam. The Gold Ridge dam recently overflowed uncontrollably after heavy rain. In 2014 and 2015 the catastrophic flooding brought by Tropical Cyclone ITA (http://www.unocha. org/top-stories/all-stories/solomon-islands-worst-flooding-history) and a close call raised alarm at various levels. 8000 people live downstream. Tens of millions of litres of contaminated water reportedly escaped from the dam via its weir.[1] That event altered the life of the 8000 downstream residents. Indeed, health authorities in the Solomon Islands released a statement advising downstream communities not to use rivers for drinking, cooking or bathing.

The environmentalist point of view was obviously that the dam did not protect the environment from the chemical contained in the tailings reservoir, therefore the tailings dam has "failed" because of the weir. For the dam engineer the weir and its associated spillway relieved the pressure and reduced the likelihood of a dam collapse. The system worked perfectly! For dam engineers a dam failure is what we have seen at Samarco or Mount Polley (see respectively Sects. 2.1 and 2.2). The weir functioning was a success from the dam engineer point of view.

In the case of the Solomon Islands' Gold Ridge dam it is obvious that a performance/success criteria should have been established for the level of contaminants under normal storage conditions such that any discharged water through the weir would meet "acceptable" levels. However, could a failure criteria be the exceedance of water quality objectives at a specified point, i.e., just before the nearest drinking water source downstream?

This leads to the question: what are the criteria for defining a tailings dam structural failure?

For example has the dam "failed" if:

1. It continues to contain all material despite being damaged?
2. It contains spills, but all solid and water impacts are limited to

   a. within property limits?
   b. within an acceptable inundation or exceedance areas?

---

[1] A weir is a structure built across a water course or on top of a dam allowing water to flow steadily over its top. Weirs built on top (at the crest) of a dam are there to allow excess water accumulated behind the dam to run away. That is, of course, without damaging the dam by uncontrolled over-topping.

3. It impacts areas beyond the acceptable inundation line?

The Canadian Cyanide Code (ICMI 2018), for example, has been effective, by requiring its members to adopt treatment to reduce cyanide levels within the pond in order to reduce bird mortality and, more importantly, to reduce the practice of ignoring freeboard limits for the purpose of meeting contaminant discharge limits. Thus it indirectly reduces the likelihood of cascading and multiple cause events. In any event, criteria for success have to be established for the mill operation including contaminant controls.

The failure criteria and performance criteria are linked. Failure stems out of performances not being attained. It is a combination of them made with "AND" and "OR" statements. The probability of the failure criteria to occur is the failure of the system p_f with consequences C evaluated by combining the various dimensions of C. The risk assessment's dimensions are, by definition, the dimensions of C.

The idea that "Processes are in place to recognize and respond to impending failure of TSFs and mitigate the potential impacts arising from a potentially catastrophic failure" (ICMM Position 2016) obviously requires clear identification and communication of success criteria and their opposite: the failure criteria.

Because consequences are multidimensional, the performance of the tailings system should encompass all of them. Thus the successful performance of a tailings systems could be declared, for example, if:

- water in the pond will be constantly maintained at least at x m from the dam;
- Seepage of certain characteristics is no larger than Q m^3/yr;
- Discharge through a weir of water with certain characteristics does not exceed q m^3/event and v m^3/yr;
- Loss of tailings through breach, pipelines' failures, etc. will not exceed k m^3/event and runout will not ....and ....;
- etc.

The points above constitute an example of success criteria for the system once all the missing parameters x, Q, q, v etc. are defined. A successful system is one that can be considered "unfailed" based on all the assumptions and conditions used through its evaluation.

The success/failure criteria are also necessary to ensure "Internal and external review and assurance processes are in place so that controls for TSF risks can be comprehensively assessed and continually improved (https://www.riskope.com/2013/11/07/social-acceptability-criteria-winning-back-public-trust-require-drastic-overhaul-of-risk-assessments-common-practice/)".

It is important to distinguish between the performance criteria (of a project, operation, company) and the success/failure criteria to be used in the risk assessment. The latter are "observer-dependent".

The performance criteria is the set of criteria for which the system is designed/created and does not necessarily correspond to the success/failure criteria necessary for the risk assessment. When performing a risk assessment it is indeed paramount to understand the metric (the "viewing angle", e.g., corporate, investor, regulators, public) of the success/failure criteria. For instance, production, maintenance, energy use, business interruption, health and safety, environmental, legal and social impacts, share value, financials, etc., can all be valid viewing angles (failure "dimensions"), but for a specific risk assessment addressing the concerns of a particular stakeholder the success/failure criteria may differ from the corporate performance criteria. This is why it is important that the hazard and risk register allows drilling from different angles to evaluate risks.

7.8 Do Not Use Common Practice Matrix Approaches, PIGs, FMEAs

The prior chapters have already shown the limitations of these common practice approaches and their bad use, and the specs have clearly shown that they simply do not make the cut anymore.

A recent paper entitled "The Risk of Using Risk Matrices" (Thomas et al. 2014) shows that oil and gas are also victims of risk matrix. The paper analyses the widespread use of Probability Impact Graphs (PIGs, risk matrix) in oil and gas. It comes to very similar conclusions to those reached by other academics and practitioners and ourselves.

However, we are very aware that most companies, dam owners and operators may already have PIGs, risk matrices (indexed or qualitative), etc., at hand.

7.8.1 "Classic" Risk Matrices Deficiencies

Risk-Acceptance Inconsistency
The regions depicted in common practice PIGs feature arbitrary stepped borders which generally bear no relation with actual corporate or societal risk tolerance. Thus those risk matrices lead to misleading decision-making support.

Range Compression
Risk Matrix and their probabilities, Consequences (Losses), risk indexes using scores to mimic expected-loss calculations do not reflect actual "distances between risks", i.e., specifically, the difference in their expected loss. Below you can see Consequences (Losses) split into five classes. Million dollars are used as a metric for Losses. In the usual log-scale (Fig. 7.4a) the intervals seem equally spaced. However, when displaying the same values in decimal scale (Fig. 7.4b) we can see a significant "range compression".

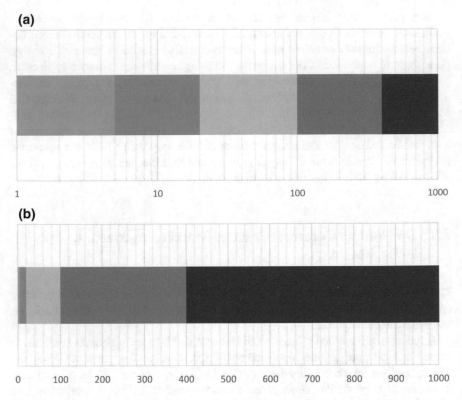

Fig. 7.4 **a** Losses classes in M$ display a somewhat uniform width in log scale. **b** Once the same classes are displayed in decimal scale the range compression phenomenon (for the lower classes) becomes evident

Users generally avoid extreme values or statements if they can. For example, if a score range goes from 1 to 5, many will select values in the 2–4 range as a result of the "centering bias".

Category-Definition Bias

Users oftentimes confuse frequency and probability. The confusion is not solved by the risk matrix compilation guidelines generally offered to users. The results are confused communications, as pointed out by Canadian CEOs (https://www.riskope.com/2016/03/23/canadian-ceos-are-looking-to-improve-measurement-risks-communication/). Researchers even talk about the "illusion of communication" generated by the "category-definition bias".

7.8.2 *"Newly Recognized" Risk Matrices Deficiencies*

Ranking is Arbitrary

If the risk matrix uses an index approach (ranks from 1 to *n*), then indexes can be in ascending or descending order. Both approaches are used in various industries.

This adds to risk-matrix-generated confusion. Ascending or, even worse, descending indexes increase the distance between a matrix-based risk assessment and reality. Rankings are arbitrary and misleading.

Instability Because of Categorization
Categories are the cardinal (index) given to probabilities and consequences ranges. Categories generate "instability" as they influence the "binning exercise" necessary to compile a risk matrix. Arbitrary range limits and categories are not necessary, as we have shown in Ten Rules for Preparing Sensible Risk Assessments (http://www.riskope.com/wp-content/uploads/2016/11/Ten-Rules-For-Sensible-Risk-Assessments.pdf).

Relative Distance is Distorted
The fact that risk-matrix axes display, explicitly or not, values in logarithmic scale distorts the distance between points. This is a phenomenon that occurs in many instances in science and economy. Unless properly explained and conveyed to users, this distance distortion leads to significant misunderstandings. Beyond the publications cited above we discussed this theme in detail in a 2013 blogpost (https://www.riskope.com/2013/11/14/a-modern-risk-manager-and-an-engineer-have-a-heated-discussion-about-fmeapigs-common-practices-and-what-a-rational-and-transparent-quantitative-risk-assessment-should-encompass/).

7.8.3   Can We Solve the Problem of Risk Matrices Deficiencies?

We show two ways:

Correcting a Series of Common Mistakes
A 2016 paper (Oboni et al. 2016) explores the idea that correcting a series of common mistakes, which can be done quite easily and inexpensively, would lead to an honest and more representative way to assess risks.

Housekeeping of Risk Register
In many instances clients call us for help with meaningless third-party risk matrices. We define the cause of that call as the "two hundred yellow syndrome". Indeed, the risk matrix's numerous deficiencies lead to many risks being binned in the central, generally "yellow", area of the matrix.

Furthermore, confusion in the hazards, consequences and risk scenario descriptions leads to added confusion. The "housekeeping of the risk register" consist in eliminating confusion by first strictly abiding to a well-defined glossary (https://www.riskope.com/knowledge-centre/tool-box/glossary/). Once the register is consistent, we vet the risks one by one, look at possible common cause failures, interdependencies and rationalize probabilities and consequences. The final result is a p-C graph unencumbered by arbitrary cell limits.

7.8.4 Bad News

The bad news is that, even after a good housekeeping, if you want to be sustainable, CSR, SLO and profitable, you will have to allow the investment to arrive to the twenty-first-century standard.

References

[BC AG 2016] Auditor General of British Columbia (2016) An Audit of Compliance and Enforcement of the Mining Sector http://www.bcauditor.com/pubs/2016/audit-compliance-and-enforcement-mining-sector

Caldwell J, Oboni F, Oboni C (2015) Tailings Facility Failures in 2014 and an Update on Failure Statistics, Tailings and Mine Waste 2015, Vancouver, Canada, October 25–28, 2015 https://open.library.ubc.ca/media/download/pdf/59368/1.0320843/5

Freeze, R. A. (2000) The Environmental Pendulum: A Quest for the Truth about Toxic Chemicals, Human Health, and Environmental Protection, University of California Press, Apr. 12, 2000

[ICMI 2018] International Cyanide Management Institute (2018) Auditor Guidance for Use of the Mining Operations Verification Protocol (February 2018) https://www.cyanidecode.org/sites/default/files/pdf/16_AuditorGuidanceforMines_2-2018.pdf

[ICMM Position 2016] International Council on Mining & Metals (2016) Position statement on preventing catastrophic failure of tailings storage facilities. https://www.icmm.com/website/publications/pdfs/commitments/2016_icmm-ps_tailings-governance.pdf

Luft J, Ingham, H (1955) The Johari window, a graphic model of interpersonal awareness, In: Proceedings of the western training laboratory in group development, Los Angeles, University of California

[MVRB 2013] MacKenzie Valley Review Board (2013) Report of Environmental Assessment and Reasons for Decision Giant Mine Remediation Project http://reviewboard.ca/upload/project_document/EA0809-001_Giant_Report_of_Environmental_Assessment_June_20_2013.PDF

Oboni F, Oboni C (2016) The Long Shadow of Human-Generated Geohazards: Risks and Crises, in: Geohazards Caused by Human Activity, ed. Arvin Farid, InTechOpen, ISBN 978-953-51-2802-1, Print ISBN 978-953-51-2801-4

Oboni C, Oboni F (2018) Geoethical consensus building through independent risk assessments. Resources for Future Generations 2018 (RFG2018), Vancouver BC, June 16–21, 2018

[Oboni et al. 2016] Oboni F, Caldwell J, Oboni C (2016) Ten Rules For Preparing Sensible Risk Assessments, Risk and Resilience 2016, Vancouver Canada, November 13–16, 2016 https://www.riskope.com/wp-content/uploads/2016/11/Ten-Rules-For-Sensible-Risk-Assessments.pdf

Renn O, Klinke A (2004) Systemic risks: a new challenge for risk management. EMBO Rep. 2004 Oct; 5 (Suppl 1): 41–S46. https://doi.org/10.1038/sj.embor.7400227. https://www.ncbi.nlm.nih.gov/pmc/articles/PMC1299208/

Roche C, Thygesen K, Baker, E (eds.) (2017) Mine Tailings Storage: Safety Is No Accident. A UNEP Rapid Response Assessment. United Nations Environment Programme and GRID-Arendal, Nairobi and Arendal, ISBN: 978-82-7701-170-7 http://www.grida.no/publications/383

Thomas P, Bratvold RB, Bickel JE (2014) The Risk of Using Risk Matrices, SPE, Economics & Management 6(2): 56–66

Chapter 8
System Definition

This chapter, System definition shows how to dissect a mining system into macro-elements, elements which are amenable to analyses.

The system definition encompasses the definition of the "business" as well as the "physical" aspects, as a sort of "software" and "hardware" taxonomy. It also requires a discussion of conflict of interest and other systemic inter-dependencies. Let's start with the "software" side with the responsibilities (Sect. 8.1) and then look at the "hardware" side (Sect. 8.2), i.e., the definition of the physical system.

8.1 Definition of the System of Responsibilities

Tailings Responsibility in situations where the regulatory system lacks substance may be dealt internally, within the mining company's system. Multinational companies have realized the need to reduce their reliance on external services and standards. They have therefore taken on more responsibility internally. Smaller companies, with fewer resources, are more dependent on outside help in developing and implementing high standards. Both ways may lead to unrealistic risk assessments (https://www.riskope.com/2017/12/06/unrealistic-risk-assessment-describing-a-rosy-scenario/) if common practices are used.

Whatever the situation, a complex relationship exists between mining companies, regulators and consultants with regard to ensuring that the highest standards of tailings dam risk assessment, design and management are identified and implemented. Many parts have to come together within a framework that examines the interrelated nature of the activities that each must perform in meeting their respective responsibilities (https://www.riskope.com/2017/11/01/unep-tailings-dams-report-and-residual-risk-assessment/).

© Springer Nature Switzerland AG 2020
F. Oboni and C. Oboni, *Tailings Dam Management for the Twenty-First Century*,
https://doi.org/10.1007/978-3-030-19447-5_8

8.1.1 Mining Company Responsibilities

Expertise in subjects such as risk assessment and emergency preparedness are often sought internally. Indeed many companies and design groups consider it desirable to have in-house experience to identify concerns in such areas and recommend that they be addressed by management. Very few recognize that performing a risk assessment is a profession requiring specific skills. Furthermore, internally-developed risk assessment may be hazardous, as it can be fraught with conflict of interest and personal pressures.

Independent Tailings Board (ITB)
An ITB should be very carefully selected, with a truly independent risk specialist and comprised of recognized industry geotechnical and management experts in the design and operation of TSFs. The ITB has the purpose of providing a company with an annual assessment of the effectiveness of its tailings governance and responsibility framework and to offer its advice and comments on key matters such as the integrity of the dam structure, the identification and management of its hazards and risks, the comprehensiveness of the assurance programs and the scope, depth and team qualifications for individual assurance activities. The ITB should also be a line of defence against normalization of deviance (https://www.riskope.com/2017/05/03/oroville-dam-risks-became-unmanageable/) and the lack of systemic resilience (https://www.riskope.com/2017/08/08/10-commandments-for-resilient-design/).

Qualified Professional Engineer (QPE)
The selection of the QPE or engineering firm for the design of a TSF should be based on their qualifications, availability and local knowledge. Furthermore, because of the complexity of tailings dams, a company needs to assess not only the lead engineer but also the composition and members of the design team. Companies should use their ITB to assist in the selection of the design consultant. In cases when the ITB may not be fully formed, the early appointment of its chair would add significant value to the selection process.

Here too the utmost care must be exercised to avoid any chance of conflict of interest, human relation difficulties, "good old boy" networks, and complacency-based behaviours.

Engineer of Record (EOR)
Leading practice requires the appointment of an EOR to have professional responsibility for the design of and changes to each TSF, to conduct annual validations of performance and to participate, as appropriate, in activities related to risk assessments, critical control measures and procedures, assurance activities.

To a great extent, the quality of a Tailings Management System (TMS) will depend on the degree of input provided by a dam's EOR and other experts on specific issues. A mine's management system only provides the framework for the documentation of the procedures required by the experts to do the job properly.

Corporate law requires directors to use their skill and experience to provide oversight of the business of a company. Directors have a duty to act honestly and in good faith with a view to the best interests of the company and to exercise the care, diligence and skill that a reasonably prudent person would in comparable circumstances. Duty of care responsibilities now requires the company's directors to provide oversight of the material risks of a corporation. Oversight of tailings dam risks is typically assigned to the board committee mandated to oversee sustainability issues.

Mining companies must accept full responsibility for the location, design, construction, operation, decommissioning and closure of tailings facilities in a manner that ensures its risk management strategies provide an acceptable level of protection for the safety, health, and welfare of the public and the environment. They should discharge this responsibility and express their commitment to the highest standards of risk mitigation through the adoption of strong policy statements based on rational risk assessments, the establishment of a comprehensive governance framework and the implementation of comprehensive assurance activities.

Finally, it may be a company's objective to ensure dams are designed and operated to the highest standards but it is impossible for a company to ensure that its facilities will be risk free.

To develop robust risk management systems and processes to identify and mitigate material risks, some mining companies have adopted ISO 31000:2009 Risk Management—Principles and Guidelines (ISO 31000, see Sect. 14.1) to provide a framework for the management of all corporate risks. The introduction to the standard states:

> ...the adoption of consistent processes within a comprehensive framework can help to ensure that risk is managed effectively, efficiently and coherently across an organization. The generic approach described in this International Standard provides the principles and guidelines for managing any form of risk in a systematic, transparent and credible manner and within any scope and context. (ISO 31000:2009)

Board leadership is an absolute necessity for a company to achieve high levels of performance. A company's tailings management programs must be driven from the top. Without such support, those responsible for designing and operating a tailings facility will have more difficulty in gaining acceptance and receiving adequate resources for what needs to be studied and designed and then operated to the highest standards.

The Chief Executive Officer is appointed by a company's board of directors and is responsible for the execution of a company's strategy and policies within the limits of the CEO's delegated authority. With regard to tailings management, the CEO will be guided by corporate policies approved by the board. Corporate risk and sustainability policies will provide general guidance, but for material risks such as tailing management specific policies and standards will also be required.

To implement the requirements of the policies as they apply to tailings management, a CEO will assign responsibility for their development and implementation to key members of the company's corporate office. In a large company this may include the chief operating officer (COO), the Chief Risk Officer (CRO), the sustainability officer and the officer assigned responsibility for the internal audit function. The assignment of responsibilities on specific aspects may be a combination of individual and team efforts. However, what will make it all work will be the demonstrated commitment of the CEO and the executive team to high standards of performance at all levels of the organization and for all activities, not just tailings management.

A common corporate practice is for a company to organize its policy requirements within a tailings governance framework with the development of a management system as the core element. Other important aspects include the application of corporate risk management policies, the establishment of organizational, design and operating standards and requirements for assurance and reporting activities.

One ubiquitous enemy to rational risk management is the corporate "Silo culture" that generates sectoral risk assessments and treats the many issues as independent "verticals". In this respect, the evolution (https://www.riskope.com/2012/05/02/evolution-of-risk-management-and-risk-managers/) has been astounding, but there is still much to be done.

8.1.2 Governments Responsibilities

Governments must be responsible for the protection of employees, the public and the environment from undue impacts and risks arising out of or in connection with mining operations. They must discharge this responsibility primarily through the development of risk based laws, regulations and guidances, by granting permits on the basis of a strong regulatory framework and public consultation and being responsible for an effective compliance and enforcement regime. Risk-based laws require quantitative assessments allowing for RIDM at the country-wide scale.

To meet their responsibilities, some governments are responding with a higher level of oversight through a more rigorous permit approval process, expanded tailings management oversight, improved compliance and enforcement activities and increased transparency. Both the mining industry and the regulators have to up their game, simultaneously, so that they can keep lines of communication open. Governments should know what they can actually request from mining companies terms of advanced risk assessments, performance predictions and long-term assurance based on uncertainty management.

8.1.3 Professional Geotechnical Engineers Responsibilities

Professional geotechnical engineers should be responsible for the design of tailings facilities in accordance with the highest state of practice and applicable regulations, statutes, guidelines, codes and standards while fulfilling their professional obligations that, "…hold paramount the safety, health, and welfare of the public…" (APEGBC 2014). Here again, rational and quantitative risk assessments performed by independent third parties are key to ensure the results.

8.2   Physical System Definition

Key to a world-class risk assessment endeavour, whether it is a ERM or a project, or a tailings operating facility is the proper definition of the system you are analysing (https://www.riskope.com/2015/02/19/one-of-ore-secrets-proper-definition-of-the-system-you-are-analyzing/). We all know that ISO and other International and National Risk Codes stress the fact that the context of the study—the environment in which your system operates—has to be described. However, so many times we have seen project teams and facilitators embarking in FMEAs or other risk-related endeavours without taking the time to rigorously describe the system anatomy and physiology. It may seem weird to use medical terms but let's see why they are appropriate.

8.2.1   Quick History

As discussed in various publications, blogpost (http://www.riskope.com/2013/12/05/some-buzz-words-and-their-meaning/) (or here) (http://www.riskope.com/2015/01/15/failure-modes-and-effects-analysis-fmea-risk-methodology/) most common practice tools date from the years of WWII and the 1950s.

At the beginning, only weapons and very "scary" systems were studied using those methodologies. Industry was still using the so-called "insurance gals (http://www.riskope.com/2012/05/02/evolution-of-risk-management-and-risk-managers/)", if any, to transfer risk, without any serious evaluations, to insurance companies willing to take a bet on them. Then, a series of mishaps, public outcry and political pressure, lead "risk" to become a buzz-word. Risk assessment and risk management were nice words to say, and common practice percolated down to the minimum common denominator to provide a "placebo" to society, but accidents were still occurring, failures were still called "unforeseeable", potential consequences were still looked at cursorily and in a compartmentalized way. No one was carefully describing the system's anatomy and physiology. It was the time of open risk workshops (tribal gatherings?) gaining the status of "instant risk assessment". Actually most of the time participants were able to voice concerns and fears, without having dissected the system under consideration, pretty much as we used to do in medicine before understanding anatomy and physiology. Then large scale terror acts (9-11-2001) occurred on US soil and in 2008 there was a global recession (http://www.riskope.com/2011/06/14/black-swan-mania-using-buzzwords-can-be-a-dangerous-habit/). All of a sudden new words were coined to describe what we humans already knew very well: poorly made risk assessments do not bring any value.

It was a feast of magic revival, obscurantism, denial of bad habits. All of those efforts just to hide one simple fact: unless we take the time and effort to properly define our systems, we cannot perform any serious analysis on them!

We humans talked then about systemic risk, non-functioning models, Black Swans (legitimate ones and silly ones), fragility, complexity, etc. The parallel is striking: if we do not know the anatomy and physiology of the human body, any surgery or drug will have a very poor rate of success, and can even make matters worse.

So, getting back to risk assessments:

- Is it true that our systems are complex? Yes.
- Are they fragile because of their complexity and other reasons? Yes.
- Do rare, extreme events occur? Yes.
- Do we have systemic risks in our systems? Yes.
- Is it true we can hide our head in the sand, say there is nothing we humans can do to evaluate the above, and merrily keep making the same mistakes? YES.
- Is it reasonable, socially acceptable, good for Humanity to do so? NO.

If you want to have fun for a moment, you can set up the same list of questions, replacing "system" with "human body" and "events" by "diseases". Enjoy!

By fostering a systematic analysis of system's anatomy and physiology, you will avoid most, if not all, of those pitfalls. That preliminary effort brings rationality, clarity and transparency to your endeavours. It makes risk studies scalable, flexible, adaptable to new conditions. It yields a holistic understanding of the risk landscape surrounding your operations/projects.

8.2.2 How to Dissect Your Tailings Dam System

The system (Fig. 8.1), not the dam, or, worse, a cross section of a dam, is the object of the required risk assessments. Looking at "the dam" by itself is already a major fallacy.

In fact, the system includes all pertinent inter-dependencies (physical, geographical, logical, informational) necessary to its operation or a clear delimitation of selected boundaries assumptions (Table 8.1).

Accordingly, the boundaries of the system define what is in the system, with respect to what is outside of the system. In fact, the system definition helps characterising threats-to and threats-from among system's elements. Finally, the definition of the project "context" in compliance with ISO 31000 (Sect. 14.1), including all

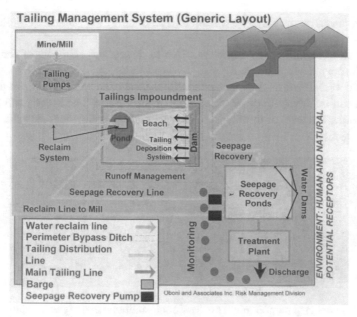

Fig. 8.1 A schematic depiction of a tailings system

Table 8.1 The first step is to define the project/operation's perimeter, the environment, personnel and cyberspace

Areas	Comments
Project/operation "perimeter"	Inside the perimeter, where operations and personnel are under corporate direct supervision and responsibility
Environment	Outside the perimeter, where external environmental damages can occur, public is present. Hazards to the system can occur in the environment
Personnel	Any person working directly or indirectly (subcontractors) for the corporation present within the perimeter. During off-hours personnel become residents of the environment
Cyber space	The realm of information, the web, data. That includes all project's/operation's IT systems, SCADA, sensors, etc. as well as the links to cyberspace (also including private cell-phones and other devices connected, temporarily or permanently, to corporate's systems)

the assumptions on the project environment, chronology etc., also helps defining the system.

With these definitions it is now possible to dissect the elements of the system within the considered perimeter.

It is important to note that elements within the system and elements in the environment can be threats-to other elements (for example, a dam breach can damage another infrastructure) or threats-from (for example, a malevolent act on a pipe by angry residents).

> Threat-to/threat-from is a type of analysis used to link identified external or internal hazards, for instance:
>
> to particular targets (elements of the system) OR
>
> from elements to targets lying outside of the system (population, environment, third parties, etc.) or inside the system (which then become interdependent).
>
> As a result each couple is qualified in terms of possible dire outcomes (consequences).

Highly-respected geotechnical professionals maintain that highest "risk" to "safe" dam operation is the mill, as the mill always underestimates the amount of water storage capacity they need for a wide variety of reasons. While directly connected, reportedly the ability to manage contaminant levels and the longer retention times required was found to be of most concern prior to the introduction of cyanide codes and guidances (ICMI 2018; Cyanide Code 2017). In order to develop a study compatible with the budgeted effort, assumptions and simplifications will generally be necessary.

The system can be described by "nodes" as shown in Fig. 8.2 where a tailings dam is split in three nodes: crown (altitude); impervious core (percolation); and upstream and downstream bodies (stability). Each node has incoming "resources", internal "processes" and outgoing "resources", and many of the resources flows can be bi-directional, i.e., affected by inter-dependencies.

Any mining system, no matter how complicated, can be described by as a system of nodes whose granularity is appropriately selected. Indeed, at the beginning the nodes may be "macro", i.e., encompass complex processes; for example, there might be a macro-node called "mill". When needed the "mill" node can be subdivided into increasing finer levels of granularity, going all the way, for example, to a node called "tailings pump A". The level of granularity is dictated by the stage of development (at pre-feasibility level only macro-nodes may be necessary) and the purpose of the risk assessment (part of the system may be modelled with macro-nodes, others be more detailed). The advantage of scalability becomes evident as the phases of development progress.

A well-defined system will avoid many blunders and confusions (https://www.riskope.com/2017/07/26/three-ways-to-enhancing-your-risk-registers/) typical of common practice risk registers.

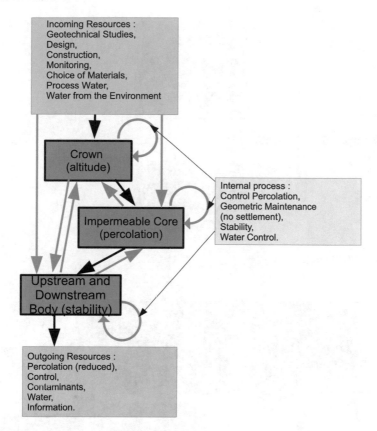

Fig. 8.2 Possible way of looking at a tailings dam as a system of three nodes

Example A: A Complex Mine with Multiple TSFs

Let's start with a single TSF with one dam in the considered mine (Fig. 8.3).

Should a risk assessment of this simple system be performed, the question would be: is Dam 15 really amenable to a significant analysis as a one element? Or should it be split into several sections because, for example, the foundation conditions vary along its layout? Or perhaps because breaches in different locations may have significantly different consequences due to land use, topographic oddities, etc.?

Figure 8.4 shows another TSF with three dams (it is perhaps actually one single dams with two corners), thus for the purpose of its risk analysis it would be preferable to split it into three elements.

Now the question is: how many sub-elements are needed to define beyond the obvious three macro-elements represented by the three connected dams? How many will capture the different nature of geology, hydro-geology and consequences?

The next step is of course to look at a complex mine with all its TSFs, ponds, main pipelines, dams and weirs. Figure 8.5 shows a real life system with a number of cascading ponds with multiple dams, weirs, independent ponds and pipelines. The

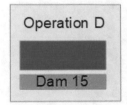

Fig. 8.3 Single TSF with one dam

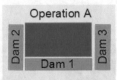

Fig. 8.4 TSF with three dams

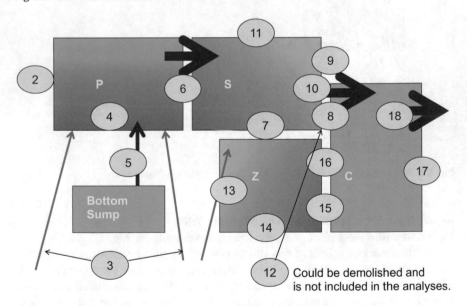

Fig. 8.5 Complex mine with all its TSFs, related dams and major pipelines, weirs

18 elements shown in Fig. 8.5 were then split into sub-elements as indicated in the previous two examples, for a total of over 50 elements to be analysed in the mine risk assessment.

Of course, there are inter-dependencies in this system (the obvious ones are those resulting from "cascading" ponds) that have to be considered in the analyses.

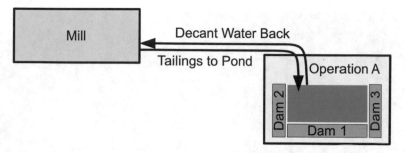

Fig. 8.6 A long tailings and decant pipelines between the mill and the TSF

Example B: A Long Tailings Pipeline Between the Mill and the TSF
The next example (Fig. 8.6) focuses on the pipelines. In this example we show tailings main and decant water, but modern mines have many other pipelines, including those for water, sea-water, and concentrates. How many sub-elements do we need to define for each pipe? Mining pipelines around the world display a huge variety of lengths, from hundreds of metres to over 100 km.

Pipelines have to be split in "homogeneous sections" as a function of their conditions (slopes, geology), environment (land use, topography, containment) and special characteristics (tunnels, river crossings, siphons, sharp bends, etc.). Potential traffic hits and fires (natural forest/brush, access) have also to be considered when splitting a pipeline into homogeneous sections.

Example C: One TSF with Diversion Ditch and Tailings Distribution Line
Figure 8.7 shows a real-life tailings pond created by a dam (sector V) and protected by a storm water diversion channel. The main tailings lines are very short, from the mill pumps to the beginning of the peripheral tailings distribution line. For the sake of the risk assessment it was necessary to split the tailings distribution line into six segments, numbered I to VI.

The rationale is to have segments that are "homogeneous" in terms of hazards and/or consequences. Hence, for example, Segment I could be hit by a diversion failure and the flooding would "immediately" report to the abutment of the dam (possible wash-out area), generating an inter-dependent failure potential. Segment II would, in contrast, possibly be hit by the diversion ditch failure, but the flooding would report to the pond centre. Segment V on top of the dam could erode the dam and wreak havoc on the structure crest, etc.

8.3 A Note on Inter-Dependencies

We just mentioned inter-dependencies again, so it is a good time to discuss this aspect a bit further. Inter-dependencies are often present even within a single dam cross-section. In this case they are called "internal inter-dependencies". Inter-dependencies are also known as "domino effects".

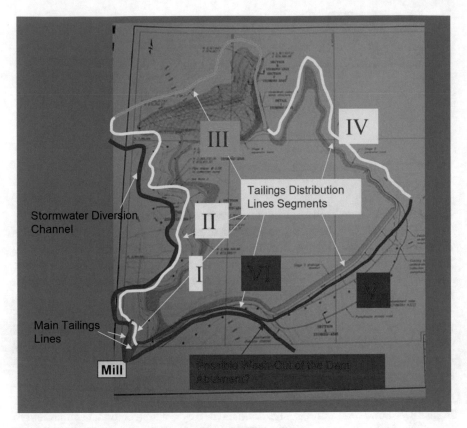

Fig. 8.7 TSF with diversion ditch and tailings distribution line

Figures 8.8 and 8.9 show two classic examples, both present in modern mines, which should be evaluated as discussed below.

8.3.1 Pipe-Dam Inter-Dependency

Figure 8.8 shows a common example of a tailings line on top of a dam. It may be exposed to hazards (see Chap. 9 for Hazard Identification) such as traffic (with construction equipment), joint failures, etc. Its probability of failure may be seen as a series system, including chances it will not be detected for a certain number of hours. At that point the domino effect can develop. If the spill can migrate toward the downstream face of the dam, erosion, then invasion of berms, then sloughing will occur. Finally, as the phenomena progress, a dam break which could probably have been reported in old statistics as due to stability will occur.

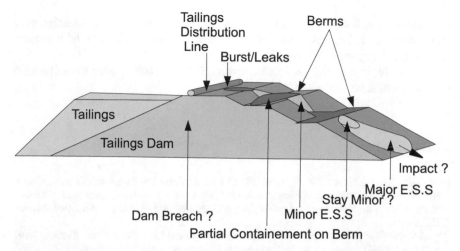

Fig. 8.8 Possible cross-section inter-dependencies. Spigotting line burst (burst tailings shown in blue) can evolve into a dam break after Erosion, Sloughing and Stability (E.S.S) start developing

What is the result of each failure?

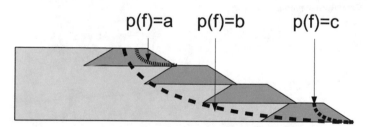

Fig. 8.9 Possible cross-section inter-dependencies. Local failures, -a- or -c-, can evolve into the global -b-

All of the above can be modelled using well known techniques such as Event Tree Analyses (ETAs) and Failure Tree Analyses (FTAs) after determining each elemental probability, for example, using the tables in Appendix A, or using loss records, or integrating any available source to derive first estimates of the elemental probabilities. Later on, there will be references and examples of this type of analyses.

8.3.2 Internal Dam Inter-Dependencies

Figure 8.9 shows a series of raises built upstream. We oftentimes see engineers evaluate the global failure (slip surface b), and local failures (slip surfaces a and c). We very seldom, if ever, see a stability analysis performed considering a domino failure, say beginning at the toe and progressing towards the crest, with the unstable mass reduced to residual strength. Another mistake is to consider the drained resistance of tailings, if they are liquefiable.

This example can also be tackled using ETAs or probabilistic slope stability analyses, which we will discuss later (Chap. 11).

References

[APEGBC 2014] APEGBC 2014. Association of Professional Engineers and Geoscientists of British Columbia (2014) Professional Practice Guidelines – Legislated. Dam Safety Reviews in BC V2.0. https://www.apeg.bc.ca/For-Members/Professional-Practice/Professional-Practice-Guidelines

[Cyanide Code 2017] International Gold Cyanide Management Code For the Manufacture, Transport, and Use of Cyanide In the Production of Gold. http://www.cyanidecode.org/

[ICMI 2018] International Cyanide Management Institute (2018) Auditor Guidance for Use of the Mining Operations Verification Protocol (February 2018) https://www.cyanidecode.org/sites/default/files/pdf/16_AuditorGuidanceforMines_2-2018.pdf

[ISO 31000:2009]. International Organization for Standardization (2009) ISO 31000, Risk Management - Principles and Guidelines, International Standards Organization https://www.iso.org/iso-31000-risk-management.html

Chapter 9
Hazard Identification

This chapter discusses hazard identification in terms of hazard vectors, scenarios, how to determine them, etc. Monitoring is discussed as well, as the mean to keep an active eye on impinging potential hazards.

Here we will discuss classic monitoring and modern techniques such as space observation. There is a note on liquefaction, which is a special hazard that cannot be ignored.

Hazards related to natural and human causes are of primary concern through all phases of the design and mine life cycle. Risks specifically related to the tailings deposition method, dam design and site selection are equally important and must be identified as part of a comprehensive approval process that is ultimately informed by a dam break analysis and inundation study. Before going any further it is necessary to go into details of the hazard definition for practical risk assessments.

When tackling a risk assessment, we define as hazard scenarios any malfunctioning or deviations from the intended level of performance of the system or any of its elements "as is", such as tailings dam failure or other phenomena, including not only those generating harm to people, properties and the environment, but also business interruption and other consequences. System "as is" means with the level of mitigation and controls present at the moment of the study and with the quality of investigations, design and maintenance which becomes apparent during the preparation of the study. Thus design choices and considerations performed by the designers/project engineers, such as:

- factors of safety (FoS);
- liquefaction (static and dynamic) potential (and the resulting requirement for undrained shear strength analyses);
- effects of reconnaissance boreholes length and number/density effects;
- etc.

© Springer Nature Switzerland AG 2020   117
F. Oboni and C. Oboni, *Tailings Dam Management for the Twenty-First Century*,
https://doi.org/10.1007/978-3-030-19447-5_9

are included in the probability of failure of each element as described later in this book, but do not constitute an hazard as defined above. The same is true for maintenance and monitoring. However, deviations from the intended level of care in maintenance and monitoring are considered as hazards.

Table 9.1 defines—as a specific example from a real-life study, not as a general list—a series of threats-from families and vectors potentially lurking on the system's elements, i.e., potentially generating hazard scenarios.

Hazard scenarios can be "elemental" or developed in long chains of "compounded/domino failures" (inter-dependencies) leading to outcomes (consequences) of an impact larger than the initiating event. In order to enable proper consideration in a rational and transparent analysis, hazard scenarios must be:

- identified and classified in a proper Hazard/Risk register. The detail of the register will be compatible with the level of information available at the time of the study;
- characterized by at least (relative) likelihood and magnitude (intensity);
- evaluated in terms of potential targets, vulnerabilities and finally, consequences;

Table 9.1 Specific example from a real-life study

Threat from	Code	Vectors
Geosphere (including climate change)	G1	Earthquake
	G2	High wind (windstorm and hurricane)
	G3	Lightning
	G4	Snowstorms (snowmelts is included in G6)
	G5	Extreme cold, freezing rain
	G6	Flooding, extreme rains/extreme drought
Electrical	E1	System of communication, IT
	E2	Power electric
	E3	Power hydrocarbons
Fixed and moving equipment	F1	Equipment failure from natural cause such as "wear"
	F2	Fire, explosion due to any hazard BUT terrorism, pandemic, outages and natural disasters
Personnel	P1	Human error
	P2	Succession planning of key personnel
	P3	Pandemic
	P4	Employee dishonesty
Public and public agents	A1	Riots
	A2	Arson
	A3	Cyber attacks

- used to generate risk scenarios which have to be evaluated, ranked in the Hazard/Risk register and properly communicated to various audiences linked to the project to allow preliminary decision making.

Provided the steps above are developed in a consistent scientific way, risks can then be compared to a risk tolerance criteria leading to rational and transparent risk ranking, but even more importantly to performing risk-based decision making (RBDM) or risk-informed decision making (RIDM) which can be understood and shared with all the project's stakeholder.

Table 9.2 displays the various "families" of elements with the potentially impinging hazards (not necessarily all apply to the specific case), scenarios, related effects and mechanisms, and finally their possible evolution to ultimate failure.

Hazard scenarios linked to information technology (IT) malfunctioning (for example sensors or monitoring delivering erroneous information for technical/intrinsic reasons) are included in the line to which they pertain for each element as IT. Hazards linked to malevolent or criminal hacks on IT systems are included in the line to which they pertain for each element as "Cyber".

In this chapter we focus on the methods to be used to gather information on the potential hazards impinging on the system's elements and sub-elements. We will close the chapter by discussing how to ensure the hazard register is maintained alive, how to check for future discrepancies, and how to decide if updates are necessary.

9.1 Standard Methods

9.1.1 Workshops, Interviews

We have seen dozens of "workshops" where a few "alpha-dogs" of any gender influenced everyone else and subordinates did not dare to talk. Furthermore, in general, these "tribal gatherings" are meaningless because the glossary is not defined or adhered to, and the system is not defined beforehand. Below is our proposed remedy to alleviate these pains.

Disruption is Key

If one watches Tim Harford's TED talk entitled "How Frustration can Make us more Creative (https://www.youtube.com/watch?v=N7wF2AdVy2Q)" a world opens up. Harford reports that psychological tests have shown that students who received handouts written with "difficult-to-read fonts" did better than students with handouts with "easy fonts"! Those tests show that "a little difficulty" leads to better results because it slows down the students and forces them to think more.

He then cites complex problem solving, which are generally considered to require a step-by-step procedure: prototype, tweak, test, improve. This widely-applied procedure leads to incremental improvements, but can also lead to painful dead-ends. If randomness is added (stupid moves, mistakes (http://www.riskope.com/2013/03/28/

Table 9.2 Various "families" of elements with the potentially impinging hazards

General element of the system	Hazard scenario	Vector, threat-from code	Effects and mechanism	Possible evolution to ultimate failure
Pipelines (tailings and dewatering (pond, polishing ponds, ditches))	Traffic (includes contractors and snow removal collisions)	G4, G5, P1	Spill from minor to full break in pipeline and other collision effects	Dam erosion/sloughing/local to general instability (including liquefaction)
	Freezing (includes deformations at flanges)	G4, G5	Spill from minor to full break in pipeline	
	Sabotage/vandalism, cyber	P4, A1, A2, A3		
	Corrosion and intrinsic mechanical failures	F1		
	Human (communication between mill and tailings operation), IT	P1, E1, F1		
Confining structures (and their content)	Geotechnical/geological and climatological, hydrological	G1, G6	Settlement, deformation, seepage/piping, high pore pressure, cracks, inadequate water balance, static liquefaction, dynamic liquefaction	Erosion, inadequate storage capacity, local to general instability (including liquefaction) flooding downstream
	Construction/building operations	P1, P3, P4		
	Operation, maintenance and monitoring during service and record keeping, IT	P1, P2, P3, E1		
	Sabotage/vandalism, cyber	A2, A3		
Outlet structures (spillways/pond to pond culverts)	Geotechnical/geological and hydrological (capacity, durability)	G1, G4, G5, G6	Unsatisfactory performance, over-topping of the crest	Related dam overtops, plug bursts, seiche/flooding downstream
	Operations (service, emergency access), IT	E1, E2, P1, P3		
	Water balance management	G2, G6	Exceeds capacity	
	Climate (includes extreme freezing)	G5	Reduced discharge	
	Sabotage/vandalism, cyber	P4, A1, A2, A3	Breaching the crest	Seiche/flooding downstream

no-failures-no-progress/)) dead-end chances are reduced and often quantum leaps occur. He adds further that if a group of professionals is tasked with solving a problem, significantly better results are obtained by a group of strangers rather than a group of friends. The perception of the friends group is that they had good time. They worked well together and performed an excellent job. In other words they are complacent.

This last statement, in particular, leads us to our work at our clients' operations. We always adopt a "Daddy what's that?" attitude. Indeed, we ask questions and never accept elliptical answers. We know by looking at the eyes of the clients' employee that after a while, this drives them nuts. However, we do not care! Our goal is to disrupt workers' complacency. We introduce that randomness that psychological tests show to be so important.

It generally works this way:

- DAY 1: we ask questions, we get elliptical answers: "I do not know"; "we have always done it that way"; "Joey knew, but is now retired (http://www.riskope.com/2012/09/27/risk-managers-are-retiring-corporations-have-to-fight-the-brain-drain-or-be-faced-with-over-exposures-to-risks/)", etc. Sometimes we actually run heavy risks of getting a punch in the nose!
- DAY 2: people know better: the disruption has provoked a tad of shame and lots of curiosity. They have discussed, possibly read old reports, called Joey… Sometimes we joke that likely the gain of knowledge on day 2 is one of the largest, immediate and most economical advantages of performing a risk assessment.
- DAY 3: interviews and meetings become more fruitful, hazard identification (http://www.riskope.com/2016/02/18/intensive-risk-management-module-at-the-university-of-turin-saa-business-school/) is easier as a creative environment is finally in place, and complacency has been mitigated.

Over the years the "Daddy what's that?" attitude has brought incredible benefits to our clients and great professional satisfaction to us.

Remember what you are doing is not an audit. It is not a policing act. You just have to be earnest in your desire to understand how the system works. What you want to do is to gain as quickly as possible an unbiased and factual understanding on how the system (operation, process, team) works in real life. You want to see what lies beyond "official" flow charts and organizational schemes.

Four "Tricks" in the Bag to Engage our Audience

(1)   The "Avatar Deck of Cards"

This deck consists of eight cards, each representing a "avatar character" typically encountered in organizations.

In Fig. 9.1, for example, you can see the card of Mrs. Rozy Scenario, avatar of the overconfident, hazard and risk-unaware character. Mr. Perryl Shield card (Fig. 9.2), is the character who believes that technological, brute force mitigation can solve any present and future problem.

Mrs. Rozy Scenario

« Noooo worries everything will be just fine as it always has...»

Fig. 9.1 Rozy Scenario avatar

Fig. 9.2 Perryl Shield avatar

Mr. Perryl Shield

« I have the RIGHT STUFF against any trouble. It's so good I am going to use it myself.»

At the beginning of the interviews, whether they are one-on-one or group, we ask participants to select the avatar they believe most closely expresses their attitude toward hazards (risks) and mitigations (http://www.riskope.com/2015/05/28/why-when-approaching-strategic-tactical-operational-planning-one-needs-to-know-about-ostriches-denial-and-prayers/). During the course of the interview, if we detect a divergence between the explanations they give and the avatar character, we keep challenging the interviewee. It is fun, people enjoy it, and it helps to find "cracks" in the stories without it being (too) personal.

(2) **The "Buddies Hazard Identification Role Play"**

We ask all the participants to split in three to four groups "as they see fit". Generally this leads to "buddies groups (http://www.riskope.com/2009/11/12/one-world-16-common-human-traits-2/)" (possibly based on age, sex, cultural background, mindset, etc.). We task each group to perform hazard identification on the system after it has been modelled and after delivering a detailed explanation of the hazard terminology.

As each participant in each group has previously selected his/her avatar, we engage the individuals and the groups based on their statements and avatars. We also ask the groups to discuss their findings. Organizational "currents of though" emerge. Alpha representatives become visible independent of their hierarchical status. Concerns and hazards are sorted out (we are not really interested in concerns, unless they prove to have the potential to generate risk exposures).

(3)   The "Alternative Buddies Hazard Identification Role Play"

This is very similar to the previous exercise (point 2), but here delegates evaluate themselves in terms of their audacity and appetite for change using a scoring system before splitting in groups as they see fit. The rating system delivers a finer evaluation than the avatar deck of cards. It opens the door to a specific test (see point 4 below) that is run during the second part of the hazard identification. In some cases we may ask delegates to use the deck of cards, then auto-evaluate and discuss blatant divergences.

(4)   The "Well Balanced Groups Hazard Identification Role Play"

In order to help us in deriving a personal objective audacity and appetite for change rating and related "talents", we present a questionnaire for participants to fill in. Delegates can now discuss the gap between their self-evaluation and the test rating, in view of understanding how this can impact their hazard awareness and mind-set. We then create working groups with "equilibrated" teams (in terms of audacity and change appetite).

As in points 2 and 3 above, we now task each group with performing hazard identification on the system. That is, of course, after we have modelled the system and provided a detailed explanation of what constitutes a hazard. As above, we engage the individuals and the groups, and ask the groups to launch into discussing on their findings. Organizational "currents of thought" emerge. Alpha representatives become visible independent of their hierarchical status. We sort out concerns and hazards (again, we are not really interested in concerns, unless they prove to have potential to generate risk exposures).

These four points summarize techniques that we find useful during hazard identification interviews. One does not need to perform each the four approaches. The selection is made on the spot, based on the preliminary interactions with the participants and the degree of willingness to participate.

9.1.2   Monitoring

Standard methods to acquire information on a dam system include the whole set of geological, climatological, meteorological, geotechnical and environmental techniques. The mere listing of them would be out of our scope here, as would providing any details related to them.

When performing a comprehensive risk assessment it is fairly common to have various information sources for data, for example, not in any specific order:

- classic weather, ground-water, topographic, deformation instrumentation, manual, or automated to some level;
- drones and/or satellite imagery and data;
- reports;
- analyses;
- documentation from engineering/technical parties, but also:
- people who have worked for many years on site, and finally:
- traditional   knowledge   (https://www.riskope.com/2017/06/07/integrating-traditionalknowledge-risk-assessments/) from local population.

Each of these comes with their own level of credibility (https://www.riskope.com/2017/05/24/big-data-or-thick-data-two-faces-of-a-coin/)/uncertainty and are generally geared toward providing information to various stakeholders, but not the risk assessor. Thus they rarely inform directly on potential failure modes.

Additionally, we would like to focus on three alarming aspects we have noted in recent decades and failure case histories:

- The tendency to rely on geotechnical investigations and instrumentation far too shallow with respect to the magnitude of dams and potential consequences of their failure.
- The fact that geotechnical investigations and instrumentation are often (mis)placed in areas that make sense from a construction point of view, but bear little significance with respect to failure modes.
- The excessive trust and reliance on the result of sophisticated software and the simultaneous abandonment of the "art of thinking" which dictated "design first a beautiful structure, think about its future behaviour, THEN calculate it".

9.2   Modern Methods

In this section we focus on satellite observation (generally called "space observation") as it makes it possible to envision large portfolio analyses (large mining companies with many dams scattered around vast territories and countries, or governments willing to keep the dam portfolio under their jurisdiction under check). Drone observation is more limited in scope and coverage, but has become so common that even magazines have presented articles on the subject (https://thebossmagazine.com/mining-with-drones/).

The information on satellite space observation and related techniques has been discussed in prior publications (Oboni et al. 2018, 2019) and taught in a course by Riskope and MDA (Oboni and MDA 2018).

9.2.1 Satellites

Data and imagery derived from space observation can be used for a variety of tailings dams observations (Oboni et al. 2018), for example to identify:

- where there are potential areas of land-use change or encroachment;
- deformations and topographic changes;
- quantitative differences in soil wetness or standing water;
- vegetation health issues relative to similar vegetation of past years.

It generally involves a manual search by an experienced interpreter of high-resolution imagery to identify the nature of the change, and to conduct a checklist search to identify any other issues of concern. The automated and manual interpretations are intended to complement each other and limit residual risk related to this element of monitoring. The automated methods are effective in drawing the interpreter's eyes to areas that might otherwise be overlooked, and the manual inspection ensures that items that may have been missed by the algorithms are caught.

Obviously the link between space observation and quantitative risk assessments is beneficial, insofar as it makes it possible to feed enhanced data into an a priori risk assessment and deliver on a regular basis updated risk assessments with a significant economy of means while answering modern requirements.

Available space observation relating to the monitoring of tailings dams includes Synthetic Aperture Radar (SAR), Interferometric SAR (InSAR), and specialized satellite sensors supplying high-resolution imagery in the visible and near-visible wavelengths. Products available in the market place that relate these observations to applications include MDA's InSAR ground movement and Radiant Solutions' PCM® technology for deriving permanent change from optical sensors. A combination of these technologies is being used to confirm and bolster data from on-site geotechnical and environmental instrumentation at various sites around the world.

Radar Observations

The use of SAR to study the earth's surface was made popular by the launch of the Earth Resources Satellite (ERS-1) and RADARSAT in 1990 and 1991, respectively. The current generation of satellites, including RADARSAT-2 can be programmed to provide a robust set of data over a mine site with observations that are not dependent on cloud cover. InSAR solutions derive ground movement from a precise observation of the time that it takes for a radar pulse to be returned to the satellite sensor.

Radar reflections (Ulaby et al. 1986) are the result of the type of surface and the geometry of the observation. For instance, flat calm water will redirect the radiation away from the radar and an image will show low backscatter values. A metal roof, oriented to reflect radiation toward the radar, is a very bright target while a ceramic structure would be basically invisible to a radar. With the understanding of the reflective properties of the structures to be observed, the dependability of radar means that it is possible to make series of observations over long periods of time that allow the automated detection of changes from, for instance, construction, soil wetness or standing water, and vegetation health issues relative to year-by-year changes.

Reproducible measurements of ground movement have been demonstrated to within 2 mm/month (Henschel and Lehrbass 2011; Henschel et al. 2015; Mäki-taavola et al. 2016). As an example, the movement on a tailings dam over a period of years is shown in Fig. 9.3. The graph demonstrates a small range of movement over a long period of time. The accuracy of the InSAR and the consistency of radar measurements make it possible to watch for the development of movement trends altering previous risk estimates. The InSAR is particularly useful for the development of baseline movement before storage or mining activities begin, i.e., at a prefeasibility risk assessment level.

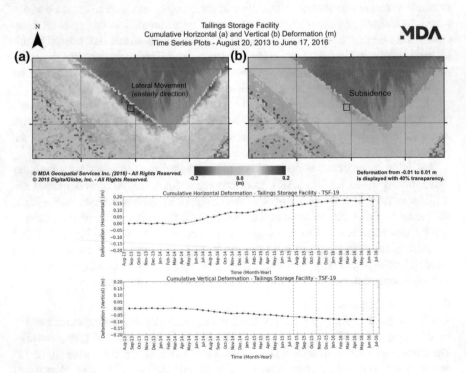

Fig. 9.3 Example of movement seen on a tailings dam by the long-term use of InSAR. The movement is shown both as a cumulative image and as a progression of the labelled square over time

Figure 9.4 (left) depicts a raw InSAR image of a mining dump impinging on a pond (black rectangle). Figure 9.4 (right) shows the topographic surface variations over a period of two months. In this case a deformation of 25 mm over two months was picked-up by means of space observation in an area where no other instrumentation was present. Based on experience, it is obvious that even the most skilled inspector would have missed that deformation during a standard inspection, should successive inspections have occurred at the same frequency. In a modern quantitative risk assessment (QRA) that deformation impacts a number of diagnostic nodes (see Chap. 11, Sect. 11.2.3) and alters the probability of failure of that slope, possibly triggering an early alert.

Optical Observations
There are a multitude of optical satellite systems available for monitoring large areas and small scales to very high resolution. From the Landsat series of satellites supported by the United States Geological Service to the high resolution sensors from Maxar's Digital Globe, optical satellites provide images of the planet in visible and

Fig. 9.4 Left: raw InSAR image of a dump impinging on a pond. Right: the ellipse encircles a zone of vertical settlement (0–25 mm) which appeared over a two-month period. © 2018 Maxar Technologies Ltd., and third parties whose content has been used by permission. All rights reserved. RADARSAT-2 Data and Products © Maxar Technologies Ltd. (2007–2018). All Rights Reserved. RADARSAT is an official mark of the Canadian Space Agency

near-visible frequencies. Visible pictures of the surface are available with resolutions of a few centimeters and can provide instantaneous descriptions of a tailings dam and its surroundings.

Optical satellites can be very powerful. The immediate recognition of objects from the imagery helps the human brain to provide context and quickly exploit the image information. Understanding change can be very quick from one image to the next but understanding the difference between a persistent change or a particular type of change can be more challenging.

Two techniques, PCM® and Intermittent Water Analysis (IW), help with the automated understanding of the magnitude and persistence of infrastructure or land use change and with the measurement of standing water cover and periodic surface water changes.

Persistent Change Monitoring

Persistent Change Monitoring (PCM®) is an automated detection algorithm that identifies multi-date change within a series of optical satellite imagery. Some of the changes that PCM® detected in real-life studies include: significant slumps or block (earth) movement; new standing water; new encroaching infrastructure such as roads, buildings, power lines or pipelines; de-vegetation and re-vegetation (from barren). Based on a process patented by Maxar Technologies, the algorithm is highly accurate and not affected by temporary features such as clouds, crop changes, or data gaps. It is now being used by commercial companies, government agencies, and organizations for a variety of applications, such as identifying areas of potential encroachment risk/features and map updating.

PCM® requires a stack of imagery to initially establish baseline conditions, followed by a series of observations over the evaluation period. Reporting includes an indication of areas where change is verified (due to its persistence over time) and also an indication where there has been potential change that has not yet been verified. It requires imagery that is automatically tightly co-registered and, as a result, is limited to medium-resolution imagery (5–30 m), such as LANDSAT or Sentinel satellites. Pending suitable image availability, PCM® provides information on historic change, which is a useful input for predictive risk assessment.

PCM® delivers data useful for quantitative risk analysis such as:

- When and where did historic change first or most recently occur?
- Which change is definite versus less likely based on the strength of the spectral difference and other factors?
- What is the probable type of change?

That information can be directly used within a risk assessment procedure to update the probabilities and consequences values.

Figure 9.5 displays an example of PCM® outputs in the vicinity of the Jim Bridger Power Plant in Wyoming. It shows two dates of imagery indicating change near the dams. The containment areas behind some of these dams have been alternately filled with water and sediment over time.

Figure 9.6 shows the PCM®-confirmed change results over that region.

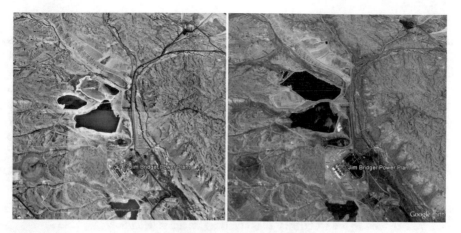

Fig. 9.5 Two data comparison of dam-related changes near the Jim Bridger Power Plant in Wyoming. © Maxar Technologies Ltd. (2018)—All Rights Reserved

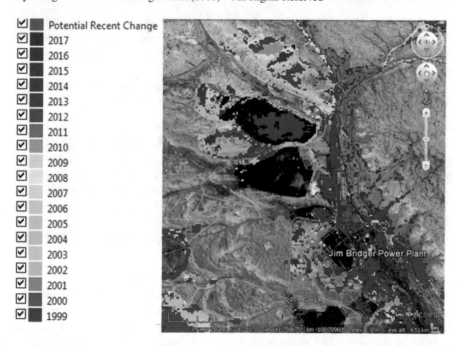

Fig. 9.6 PCM® result for confirmed change. © Maxar Technologies Ltd. (2018)—All Rights Reserved

Intermittent Water Analysis

Intermittent Water Analysis (IW) layers characterize water extent on the Earth's landscape over time. IW products are derived from a stack of over thirty years of LANDSAT imagery, often totalling upwards of 500 historical images per location. A sophisticated process for very accurately extracting water at the pixel level is used to map water extent on each image individually. The individual water masks are then stacked and analysed to count the number of water observations and frequency at each pixel. Frequency is summarized on the full dataset, decadal subsets of the images, and monthly subsets of the images. Finally, the frequency derived from the full set of imagery is used to generate a vector geodatabase of the naturally occurring lakes and ponds in the footprint.

There are three general categories of deliverable layers for each 30 m pixel:

- layers related to the frequency that standing water is detected;
- layers related to the frequency that snow or ice is detected;
- relative soil moisture.

Indeed, the space observation techniques described above seamlessly integrate with modern risk approaches, supporting tailings dams management for the twenty-first century for medium to large dam portfolios, where the costs of traditional monitoring techniques would be prohibitive.

Each of the categories above has additional statistical layers related to observations clustered by month or decade, as well as average or extreme values per pixel.

- Historical InSAR Deformation Analysis: analysis covering historical observation period (depends on extant coverage) using archived RADARSAT-1 satellite images designed to establish historical deformation trends within the area of interest.
- Forward InSAR Deformation Monitoring Program: this monitoring can be deployed for a selected observation period (for example, one year) using RADARSAT-2 satellite images.
- Persistent Change Monitoring (PCM®) Analysis.
- Intermittent Water Analysis (IW) layers characterize water extent on the Earth's landscape over time.

The techniques lead to probabilistic analysis of dams and dykes with an automated or semi-automated probability updating system, based on the dynamic link between space observation and QRA mentioned earlier. Of course, like all systems of this kind, caution will have to be exerted during all phases of the deployment. This is not a universal panacea, but, to use an automotive metaphor, it provides a good set of lights to drive through the night and a couple more instruments on the dashboard to alert the driver in case of an emerging problem, which is an impartial, fact-driven, emotionless aid to driving. Thus, if we cannot ensure that sudden, unforeseeable failures will never occur, we can certainly say that the results brought by this approach will enhance planning and mitigation capabilities. It is indeed possible to develop probabilistic updating of various types of data which may include, among others, deformation velocity (for example, cm/year), and number of events of a certain magnitude (for example, number of events exceeding a certain magnitude per year), etc.

The updating then makes it possible to re-frame probabilities present in the quantitative risk register and to re-evaluate the risks.

The various examples above show the benefits found in linking multi-temporal objective space observation with a dynamic convergent QRA platform in mining projects and operations. The two-pronged approach enables us to "measure" and make sense of a complex problem. It allows us to:

- transparently compare alternatives;
- discuss rationally and openly the survival conditions;
- evaluate the premature failure of a structure.

Connecting a dynamic QRA platform with a high-performance data gathering technique reduces costs, avoids blunders, and constitutes a healthy management practice, especially for long-term projects requiring short- or long-term monitoring, including, of course, site restorations.

Principles for Linking Space or Drone Observation to Risk Assessments
The purpose of this section is not to deliver the details of the how-to connect space or drone observation to probabilistic analyses of dams and dykes, for obvious scope limitations. Instead, it is to explain the principle and the angle of attack leading to prepare an automated or semi-automated probability updating system, i.e., how to generate the link between space or drone observation and QRAs mentioned earlier.

Table 9.3 explains how various space or drone observation themes can be approached to deliver useful data for an a priori risk assessment. That means using a historic space or drone observation database on a site that is being assessed for risks the first time.

Let's now examine which data can be gathered on a regular basis over cycles of observation and the principles by which they can be used in updating probabilities and consequences of a quantitative risk assessment. As an example, consider a portfolio of dams or dykes. We will first review freeboard, wet spots and dam deformations, and finally symptoms-driven methodologies. We will then discuss the principles of automated re-evaluation procedures.

- It is generally agreed that the amount of freeboard of a dam or levee should be increased to protect areas with high value and high loss potential. A review of different freeboard requirements in various countries provides some examples of commonly adopted values (MacArthur and MacArthur 1991). Furthermore, the beach length (distance between the crest of the dam and the water in the tailings pond) reduction is a well know indicator of poor potential stability conditions in the dam. Both freeboard and beach length can be easily measured by means of satellite observation and have an impact on the probability of failure.
- The same occurs with wet spots on the downstream face of a dam, at the toe. Dam deformations (vertical and horizontal) can be compared with a known height/deformation ratio to determine the state of the structure.
- Finally, symptom-driven methodologies can be applied. Characteristics observable by satellite or drone which reportedly alter the probability of failure of a dyke or a dam are, for example:

Table 9.3 Space observation themes versus data for a priori risk assessment and benefits

Space observation themes	Approach	Data for a priori risk assessment	Benefits from enhanced data
Visual inspection	Visual comparison of imagery at different times	Pre-existing damages, damage evolution	Increased efficiency of site visits (if still necessary) based on history
Quantitative geometry, topography data	Contour lines comparison (volumes, deformations), tension cracks, cross sections definition	Detection of potentially unstable volumes, slow creeping volumes, deposit and erosion	Enhanced hazard identification of existing or potential phenomena such as slides, debris flows, flash-floods, rockfalls etc. Probability estimates
Quantitative surface hydrology data	Water/solid flows, drainage patterns, erosion patterns and deposit areas definition	Detection of potential debris flows, drainage malfunctions, overflows	
Quantitative slope cover data	Slopes, orientation, vegetation cover health and stress definition	Detection of vegetation distress and its causes. Permafrost loss potential. Snow accumulations	

- Existence of low spots (not enough freeboard, leading to over-topping probability increase);
- Presence of erodible material on the slopes;
- Over-steepened slopes;
- Narrow spots on the crest;
- Encroached width, toe erosion, works at toe, etc.;
- Stressed vegetation;
- Easily accessible structure, poorly secured perimeter, residences proximity;
- Lack of maintenance/repairs.

   The automated re-evaluation procedure principle makes it possible to use space or drone observation data to update probabilities of failure and consequences. As discussed earlier, a preliminary risk assessment estimates a range for the likelihood of failure of various events. The range can then be split in positive partial contributions by giving relative weights to the positive observable characteristics, based on results in the literature, for example as displayed in Table 9.4, which uses a dyke case study.

   This leads to determining the dyke's likelihood of failure and finally to the prioritization for each segment based on positive space-observable characteristics (extant mitigative measures and features) as shown in Fig. 9.7.

Table 9.4 Positive observable characteristics

	Characteristic	Relative weight (%)
1	Not a low spot	19.30
2	Riprap on the "waterside"	10.50
3	Mild pitch of the dyke on the "dry" side	6.60
4	Extra dyke width, crest width of the embankment	3.85
5	Revetment of the embankment crest (paved, etc.)	3.30
6	Encroached width, toe erosion, etc.	2.75
7	Trees on the dyke on the "waterside"	1.65
8	Easily accessible, not private property, close from residences	1.10
9	Low flow velocity at the toe of the embankment	0.55

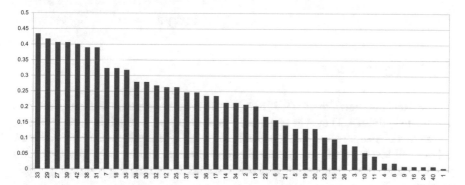

Fig. 9.7 Segment prioritization based on positive observable characteristics

We were very pleased to see InSAR being referenced in the Expert review of Cadia tailings facility (http://www.newcrest.com.au/media/market_releases/2019/Report_on_NTSF_Embankment_Slump.pdf) as we believe space observation data integrated with the methodologies explained in Chaps. 14 and 15 will help solve many problems related to tailings dams portfolios risk prioritization and mitigation roadmap.

In the case of Cadia, InSAR data were used a posteriori by the Expert reviewers, but they are nevertheless very interesting. We take from this that the InSAR measures revealed a significant increase in displacement rate in the three months prior to failure and that there was a similar trend between calculated and InSAR measured displacements.

Fig. 9.8 Big data versus thick data

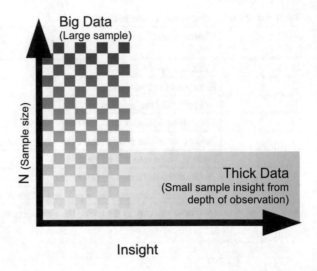

9.2.2 Big Data and IoT

Big data and internet of things (IoT) are going to become common features in all sorts of mining activities. They will help define better ranges for reliability and failure of a system's elements, and make it possible to search world-wide occurrence of near-misses and losses, news, etc. At the other end of the spectrum, thick data are useful to understand deep motivations and can foster SLO and CSR by fostering proper communication.

Big data and Thick data are actually two faces of a coin (https://www.riskope.com/2017/05/24/big-data-or-thick-data-two-faces-of-a-coin/) which can be defined as follows (Fig. 9.8):

- Big data is a term for large or complex data sets that traditional software has difficulties processing. Processing generally involves, for example, capture, storage, analysis, curation, searching, sharing, transferring, visualizing, querying, updating, etc. However, the term Big data also often refers to the use of predictive analytics, user behaviour analytics or certain other advanced data analytics methods. Analysis of data sets can find new correlations to spot business trends, prevent diseases, combat crime and so on.
- Ethnographers and anthropologists, adept at observing human behaviour and its underlying motivations, generate thick data. Thick data is qualitative information that provides insights into the everyday emotional lives of a given population.

To date, different people have supported and used big data and thick data. These are organizations grounded in the social sciences (thick data) versus corporate IT functions (big data). This constitutes a perfect example of silo culture: Big data (http://www.riskope.com/2014/05/22/big-data-and-risk-assessment/) and thick data should "talk to each other", but most of the time do not because of silo culture.

Big data relies on machine learning, isolates variables to identify patterns, reveals insight. Big data gains insight from scale of data points, but loses resolution details. It does not tell you why those patterns exist.

Thick data relies on human learning, accepts irreducible complexity, reveals social context of connections between data. Thick data gains insight from anecdotal, small sample stories, but loses scales. It tells you why, but misses identifying complex patterns.

Focusing solely on Big data can reduce the ability to imagine how the world might be evolving. Big data only is not sufficient for risk assessment, and in particular hazard identification (http://www.riskope.com/2017/04/05/hazard-identification-science-art/). It can create a distorted view of the risk landscape surrounding an entity.

Known Knowns and Unknown Knowns

If one is seeking a map of an unknown risk territory (risk landscape (http://www.riskope.com/2016/06/23/business-intelligence-platform-helping-tailings-stewardship/)) and data are scarce, then thick data is the tool of choice. As data availability grows, on its way to becoming "big data", integrating both types of data becomes important. In the case of innovative companies that combined insight can be highly inspirational.

When performing risk assessments of dams, mining operations, etc. we always collect and analyze "stories", anecdotes, loss reports to gain insights of "pre-existing" states of the system. The combined insight can tell us that a dam that "looks wonderful" actually has a "congenital defect" that raises the probability of failure.

Big data would not be capable of showing that, but could probably reveal a pattern between third-party observations and meteorology. In fact it could reveal a patterns between any other groups of variables, which could sound an alarm on "shorter-term" emergent hazards.

Integrating Big Data or Thick Data

Working successfully with integrated big and thick data certainly enhances any risk assessment. Over the years we have found ways to integrate data from multiple sources and of various natures in our risk assessments. We routinely use incomplete thick data sets in conjunction with expert opinions and literature to generate first a prior estimates of the probability of occurrence of hazards and failures. This immensely increases the value of the first-cut risk assessment, which can then be updated using big data and Bayesian techniques (http://www.riskope.com/knowledge-centre/tool-box/glossary/) (see Sect. 10.1.2).

The combined approach also makes it possible to enhance the value of big data, avoid capital squandering, and the reduce the running cost necessary to obtain big data. Recent studies have shown that without that approach data oftentimes remain virtually "unused" (https://www.inc.com/jeff-barrett/misusing-data-could-be-costing-your-business-heres-how.html).

9.2.3 Conclusion

Integrating big data and thick data brings value and should be fostered. Using big data in isolation can be problematic. Thus it is crucial to explore how big data and thick data can supplement each other. This demands the integration of qualitative evaluation and expert-based judgments with "hard" quantitative data. While we recognize that melding big and thick data together isn't easy, we do it on a daily basis.

9.3 A Note on Liquefaction: A Special Hazard

There is an important body of literature on static and dynamic liquefaction of fine sands and silt. Interested readers should probably start with an article entitled "Static Liquefaction Of Tailings – Fundamentals And Case Histories" (Davies et al. 1998).

Static liquefaction, and the resulting flowslide of liquefied tailings materials, is a relatively common phenomenon among tailings impoundment failure case histories. Static liquefaction can be a result of slope instability issues alone, or can be triggered as a result of other mechanisms. It is a special hazard as it can be both cause and consequence of failures. Here are a few examples from the 1990s:

- Sullivan Mine, Canada, 1991. The dam had been built on a foundation of older tailings that were placed as beach below water material. The failure of the upstream-constructed facility was triggered by shear stresses in excess of the shear strength in the foundation tailings. As the material strained, the pore pressures rose and drainage was impeded, leading to liquefaction event. The downstream slopes of the dam averaged roughly 3H:1V. The failure was very brittle and sand boils, water expressed from standpipes, and other "classic" liquefaction expressions were evident. The only trigger to the liquefaction failure was the slope geometry: a pre-failure dyke slope of about 2.5H to 3.5H:1V with a maximum dam height of about 25 m.
- Merriespruit Harmony Mine, South Africa, 1994. A relatively minor rainstorm caused an over-topping event, and runout caused toe erosion, which in turn initiated the flow failure. The tailings were quite fine-graded, with more than 60% finer than 74 μm. However, these fines were also essentially cohesion-less and once an area of the dam toe was eroded and local slopes were increased to the range of 2H:1V, static liquefaction and the massive flowslide was initiated soon after.
- Los Frailes Mine, Spain, 1997. The initial movement of the rock-fill dam (due to a foundation failure in a weak layer of marls) was the triggering mechanism that allowed the tailings to liquefy. This liquefaction exerted a thrust against the dam that contributed to the relatively rapid progression and large lateral displacement of the overall failure event.

As it can be seen from these three examples, liquefaction can either be the trigger of a failure (Sullivan Mine) or the consequence of another mechanism (Merriespruit, Los Frailes).

In any case, static liquefaction occurs when rapidly loaded, loose saturated sand (such as that deposited in an underwater tailings beach) do not have time to drain (dissipate the water pressure) and hence the soil collapses into a liquid form. The prediction of the in situ undrained strength for these materials is highly uncertain due to the intrinsic uncertainties on the initial void ratio and on the fabric (the way sand grains are packed together) of field scale deposits of these materials (tailings ponds).

The most readily identified of these rapid loading conditions, at least from a design perspective, is the transient loading from seismic events. Whether limited deformation or eventual flowslide development, the effects of transient seismic loads on mine tailings are well documented in the literature and well recognized by current engineering standards. However, there are many other rapid or undrained shear loads that affect mine tailings.

The numerous potential triggers of undrained response can be of equal importance to seismic loads due to their more common occurrence at mine sites. Included in these common loads are incremental raise construction and episodic spigotting. The first can lead to relatively rapid increases in stress levels and undrained conditions in susceptible materials, while the second can cause temporary changes to the amount of tailings saturated in a given section of an impoundment. Conversely, traditional static loads are taken to be those in place for a considerable period. Other mechanisms, such as a transient saturation of the downstream shell of a tailings structure, can also trigger liquefaction due to rapid reductions in effective stress.

When any combination of triggers is possible it should be automatic practice to invoke undrained strength properties (S_u) in loading situations where significant pore pressures could develop, i.e., performing undrained stress analysis (USA). Additionally, the safety of dams susceptible to static liquefaction is perhaps even better expressed in terms of the cumulative probability of potential triggering mechanisms. Despite the obvious difficulties in quantifying this cumulative probability, the approach has the virtue of at least recognizing the potential for static liquefaction.

Here is a list of well documented potential static liquefaction triggers in tailings impoundments including:

- Increased pore pressures induced by an increase in the piezometric surface, and/or change of pore pressure conditions from below hydrostatic to hydrostatic, or to higher than hydrostatic. Thus poor management/monitoring, but climate change event could also be a trigger.
- Excessive rate of loading due to rapid raising of the impoundment. Also rapid rate of rise as the trigger for an upstream beach below water leading to static liquefaction failure. Poor Management is a trigger.
- Static shear stresses in excess of the collapse surface, leading to "spontaneous" liquefaction. This is for example the result of aggressive slopes, foundations deformations, etc.
- Removal of toe support from an overtopping event, lateral erosion from a watercourse encroachment or any other situation when the toe can be steepened/removed.

- Foundation movements rapid enough to create undrained loading in tailings material susceptible to spontaneous collapse.
- Heavy equipment vibrations e.g. working on dam.
- Blasting operations in the vicinity.
- Any work at dams' toes.

References

Davies M, McRoberts E, Martin T (1998) Static liquefaction of tailings - fundamentals and case histories, 51st Canadian Geotechnical Conference, proceedings vol 1, 123–132, Canadian Geotechnical Society

Henschel, MD, Lehrbass B. (2011) Operational Validation of the Accuracy of InSAR Measurements over an Enhanced Oil Recovery Field. Proc. 'Fringe 2011 Workshop', Frascati, Italy, 19–23 September 2011 (ESA SP-697, January 2012) https://earth.esa.int/documents/10174/1573054/Operational_validation_accuracy_InSAR_measurements_enhanced_oil_recovery_field.pdf

Henschel MD, Dudley J, Lehrbass B, Shinya S, Stockel B-M (2015) Monitoring Slope Movement from Space with Robust Accuracy Assessment. Proceedings of the 2015 International Symposium on Slope Stability in Open Pit Mining and Civil Engineering, The Southern African Institute of Mining and Metallurgy, Johannesburg, pp. 151–160

MacArthur RC, MacArthur TB (1991) International Survey of Levee Freeboard Design Procedures. In: Hydraulic Engineering: Saving a Threatened Resource—In Search of Solutions, ed. Marshall Jennings, American Society of Civil Engineers, New York

Mäkitaavola K, Stöckel B-M, Sjöberg J, Hobbs S, Ekman J, Henschel MD, Wickramanayake A (2016) Application of InSAR for monitoring deformations at the Kiirunavaara mine. 3rd International Symposium on Mine Safety Science and Engineering, Montreal, August 2016 http://isms2016.proceedings.mcgill.ca/article/view/73/25

[Oboni and MDA 2018] Oboni Riskope Associates Inc., MacDonald Dettwiler and Associates, Ltd., Tailings 2.0 course: Space observation, quantitative risk assessment bring value and comply with societal demands, Tailings and Mine waste 2018, Keystone, Colorado, 2018

[Oboni et al. 2018] Oboni F, Oboni C, Morin HR, Brunke S, Dacre C (2018) Space Observation, Quantitative Risk Assessment Synergy Deliver Value to Mining Operations & Restoration, (https://www.riskope.com/wp-content/uploads/2018/06/Symposium-2018-article-MDA-riskope_2018-03-03.pdf) (see the presentation (https://www.riskope.com/wp-content/uploads/2018/06/Rouyn-Noranda_presentation-2018-06-01.pdf)) Rouyn-Noranda, 2018, Symposium on Mines and the Environment, Rouyn-Noranda, Québec, June 17–20, 2018

[Oboni et al. 2019] Oboni F, Oboni C, Morin R (2019) Innovation in Dams Screening Level Risk Assessment, International Commission on Large Dams/Commission Internationale des Grands Barrages (ICOLD/CIGB), Ottawa, Canada, June 9–14, 2019

Ulaby FT, Moore RK, Fung AK (1986) Microwave Remote Sensing: Active and Passive: Volume 2: Radar Remote Sensing and Surface Scattering and Emission Theory, Norwood MA, Artech House

Chapter 10
Defining Probabilities of Events

This chapter, Defining probabilities of events discusses how probabilities of single events can be evaluated, updated and delivers some examples of portfolio analyses.

Probabilities measure the chance an event will occur. In risk assessments and project evaluations we are not dealing with absolute probabilities, but with relative probabilities (within a portfolio) over a specific time. Generally the time is one year and we use therefore annual probabilities. Do not confuse annual probabilities and frequencies which measure average number of occurrence over a certain time interval. The confusion comes from the fact that the number expressing probability and frequency is very similar, once it goes below say 1/10. Hence a probability of 1/10 has the same number—0.1—as a frequency of 1/10, but they do not mean at all the same conceptually.

In particular, the use of frequencies expressed as one event occurrence over n years (e.g., 1 flood over 200 years) misleads people into thinking that that event will actually only occur every $n = 200$ years. Nothing could be more wrong: there are plenty of 1/200 years events that have occurred several times in a decade, for example!

Using the Poisson distribution, it is possible to link the number of occurrences of an event over a selected time t to the mean occurrence rate (frequency). For example, if 15 catastrophic tailings dam breach events over 10 years have been observed (this has occurred in the world portfolio in the recent past), that means a measured frequency of 1.5 events/year.

Using Poisson's distribution it is easy to compute and graph the probability of seeing any number of events (1, 2, ... n) during, for example, a single year. In Fig. 10.1 the vertical axis shows the annual probability and the horizontal axis the number of events. We can see that with that frequency one event per year has $p = 0.33$ to occur "next year", and three events per year has $p = 0.12$, etc. As on-site monitoring and/or space observations deliver new occurrences of events, frequency and related probabilities can be updated in a new cycle, leading to updated risks.

© Springer Nature Switzerland AG 2020

F. Oboni and C. Oboni, *Tailings Dam Management for the Twenty-First Century*,

https://doi.org/10.1007/978-3-030-19447-5_10

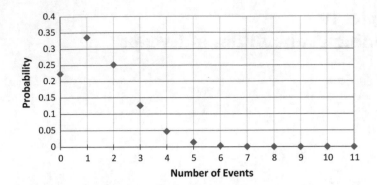

Fig. 10.1 With a measured frequency of 1.5 events/year the probability of occurrence (vertical axis) of 1, 2, 3, … n events next year (horizontal axis) can be evaluated using Poisson's distribution

For small frequencies (up to 1/10 years (units)), one can assume annual probability ≈ frequency, as shown in Fig. 10.2. However, at frequency = 1/5 years (f = 0.2) the error of the approximation rises to 20%. After that, you need Poisson's distribution.

The probability of an event occurring once or more within its return period is always 0.63. This is because the likelihood of seeing no event within its return period is 0.37, which is e^{-1}.

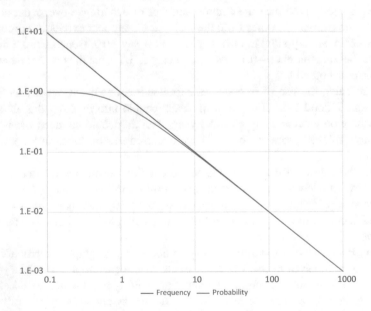

Fig. 10.2 Frequency versus probability plot. The vertical axis shows probabilities (bound to one), the horizontal axes bears frequencies

Probabilities are perishable goods. They change as the system evolves, hence they require updates. Frequencies, which are linked to "long-term averages" of occurrences vary too, but are slower to change. In mining we may well consider an horizon of two to five years and assume initially probabilities will remain "fresh" that long, but we need to keep an eye on this and perform updates as soon as needed.

TSFs oftentimes have service lives that span over multiple decades and their closure and post closure lives may very well span over respectively another set of multiple decades and centuries. There are plenty of cases where the "P" word, "perpetuity" is used. Just remember, perpetuity is way longer than a long time! Below we try to give a sense to the scale of time using an excerpt from a book chapter we recently wrote (Oboni and Oboni 2016).

> If our mining company had existed:
>
> **45000 years ago**: we would have been mining hematite at Bomvu Ridge in Swaziland. Our mine would have produced insignificant waste and no societal concerns.
>
> **3000 years ago**: we would have been bidding for the rights to mine silver at Laurentia just east of Athens. Slaves would have worked to our command to provide the money to build the Parthenon.
>
> **1000 years ago**: we would probably have been asked to sponsor the crusades. Total world population was around 300M souls. Languages used then have since disappeared.
>
> **500 years ago**: we could have learned about Mr. Columbus's recent discovery. Total world population had increased to 500 M souls.
>
> **200 years ago**: we would have been concerned by the Battle of Trafalgar, Napoleon's retreat from Moscow, and the Battle of Waterloo. The world's population had reached 950M souls.
>
> If we had closed our mining tailings facility 1000, 500, or 200 years ago, would we have expected that the tailings (mining wastes) should still be right there where we dumped them, unattended, not maintained, not monitored? Oh, we are forgetting one thing: had we left a Standard Operating Procedure and Maintenance Manual for "future generations," now the manual would be in a language difficult (if not impossible) to understand. The documents might have turned to dust or have been heavily damaged. In addition, if we think digital transcriptions of our documents may have saved us, the solar flare of 1859 (the Carrington event) would probably have erased them all if fires, floods, and wars had not destroyed them earlier.
>
> Keep this information in mind as you go through the rest of this section.

We detailed the theme of the long term survivability of TDs at TMW 2014, where attention was focused on modeling the aging process of a geo-structure as a series of discrete hits by hazardous conditions (these could be anything, from an earthquake to flooding, etc.). In Oboni et al. (2014) an attempt was made to draft a multidimensional estimate of future consequences. We stated:

> … Should the value of consequences increase… then the "excellent dam" would soon pose a societally unacceptable risk even for shorter terms. Any dam that starts its life with a small initial FoS or reduced standards of care… would see its risk evolve towards intolerable

societal risks faster, even if its consequences of failure remain constant.... The methodology
developed in this paper enables us to "measure" and give a sense to a complex problem, to
transparently compare alternatives, to discuss rationally and openly the survival conditions,
or to evaluate the premature failure of a structure. The only way to slow down the increase
of the probability of failure is to repair damage occurring as a result of each hazard hit,
or to entirely avoid the damage. The second is generally "not feasible" for economic and
constructional reasons. Risks, especially long term ones, can never be reduced to nil.

As this present book is focused on the service life of TSFs and the closure phase
but not the long-term, "perpetuity" phase, we refer interested readers to (Oboni et al.
2014).

10.1 Probabilities of One Event

The scope of this text does not include mathematical skill development. However
there are a few mathematical tools whose concepts are important in the deployment
of sensible risk management when statistics are poor or non-existent and projects
evolve for many years, decades.

Probabilities allow us to consider the various sources of uncertainty and evaluate
their impact on the big picture. Thus we are of the opinion that even rudimen-
tary probabilistic analysis (https://www.riskope.com/wp-content/uploads/2015/10/
Tailings-Facility-Failures-in-2014-and-an-Update-on-Failure-Statistics.pdf) is better than working deterministically. In fact, the inclusion of uncertainties is far superior
to "artificial" parametric studies (as, for example, varying one or two parameters at
a time to see their influence on the overall results).

The key question of what constitutes the essential (understood as basic, indis-
pensable) and ideal (understood as "perfect") data set to use is frequently asked by
users of risk assessments. There is no "simple" answer to that question, as we often
deal with facilities that may not even have been commissioned yet and past performances may not reflect future behaviour because of system or climatic changes.
Indeed, any internal or external change to the system has the potential to invalidate
the assumption that past experiences are sufficient to understand and calibrate future
implementations.

Let's also remark that no risk assessment ever has the ideal data set available.
Indeed available data are generally gathered for other purposes, may be censored
and biased, and, most importantly, they reflect the past, not the future. This is the
case even in extremely regulated environments. Therefore the analyst must rely on
his/her skill and specific knowledge to use available data, either specifically from the
site(s), or from specific technical literature, to define framing probabilities ranges.

Of course any factual data (for example, over-topping, erosion, drainage and
deformations) will help immensely in framing a reasonable range of probabilities.
Records of near-misses can also be considered essential. Accident records can be
considered essential, although in many studies done in the past, there were no such
records, simply because the facility was not even in service. For future facilities,

"essential" means "reported in the literature", and in some cases "collected expert opinion" together with an encoding methodology that makes it possible to transform "knowledge" into a probability.

10.1.1 Initial Estimates

Appendix A shows how the probability of failure of tailings dams and other events scales with respect to many real-life examples. The tables in Appendix A can be used to generate first estimates of probability of failure of various events, in relative terms, when no or very little statistical data and history are available.

By selecting a wide range of probabilities (and consequences) for each event a risk assessment will become amenable to a Bayesian update of probabilities when new data become available. Bayesian updates are developed after a first, a priori evaluation is mathematically corrected using the Bayes theorem, as new data become available. New data become available as monitoring and observations on site progresses and, in some cases, when clients ask for their operations to be monitored by space observation.

We generally start a QRA (called the a priori assessment) using uniform distributions for probabilities and consequences. The uniform distribution leads to the most conservative estimate of uncertainty, as it gives the largest standard deviation (NIST/SEMATECH 2012). The uniform distribution makes it possible to evaluate first and second moments of functions of stochastic variables "by hand" (by means of direct formulae or using the point estimate method developed by Rosenblueth (1975)), making it possible to bypass "black-box" solutions, such as the Monte Carlo simulation, which, again, give a sense of false precision. Interested readers can go deeper into the theme of using uniform distributions as a priori distribution in a Bayesian approach (see Sect. 10.1.2) by reading the literature on uninformative priors[1] (Carlin and Louis 2008).

On the other hand, in some cases—for example when adding independent random variables, or considering higher levels of information for a variable (e.g. min-max, first and second moment, i.e., average and standard deviation) based on the Central Limit Theorem—it is possible to assume that the result tends toward a normal distribution (informally, a "bell curve") even if the original variables themselves are not normally distributed. The Central Limit Theorem implies that probabilistic and statistical methods that work for normal distributions can be applicable (with caution) to many problems involving other types of distributions.

The next step, as the information level increases, is to use an empiric distribution, such as a Beta distribution. Finally, when the knowledge increases further, the "real" distribution of a variable can be determined (exponential, Gumbel, etc.). Each time

---

[1]The term "uninformative" is common but misleading, as the simple knowledge or estimate of min-max already constitutes a very valuable piece of information.

an analyst "feels the temptation" to use a higher level of distributions, perhaps to perform a Monte Carlo simulation, the trade-offs and potential for a misleading result must be carefully considered.

10.1.2   Updating Probabilities

As shown in the previous section, it is possible to develop probabilistic updating of various types of data, which may include, to mention just a few: deformation velocity (for example cm/year); number of events of a certain magnitude (for example number of events exceeding a certain magnitude per year); etc. The updating makes it possible then to re-frame probabilities present in the a quantitative risk register and to re-evaluate the risks. In what follows we present a few techniques that can be used to update probabilities.

Exceedance Probability Updates

The exceedance probability is the probability of an event being greater than or equal to a given value, i.e., that the event will exceed, for example, a given magnitude. Forecasting the future exceedance of previously observed extremes is extremely important for risk assessments. Based on repeated observations it is possible to re-frame the probabilities of exceedance and thus to rationally update the risk register.

Bayesian Updates

Bayesian analyses allow to update frequencies and probabilities as new data are generated (Ang and Tang 1975; Straub and Grêt-Regamey 2006) by space observation or other means. Consider, for example, the case where the available information is a set of observed n detached rocks from a slope, which are described by their volume and the time during which they occurred. Note that the Bayesian update will be valid only insofar as the observations are free of error (i.e., all rocks are recorded); this is the reason why regular monitoring is a necessity. In order to allow later Bayesian updates a quantitative risk register should include the a priori estimate of frequencies or probabilities. If no data are available beyond a min-max range defined by models or expert opinions, the simplest and oldest rule is to assume a uniform distribution (Fig. 10.3), as stated earlier. However, if sufficient data were available, the risk assessment could also be set-up with a more refined "prior" distribution and then use Bayes to obtain the first "posterior" distribution, the second posterior, etc. The application of Bayes shows that one single event provokes a shift of the distribution, as shown in Fig. 10.4.

10.1.3   Summary of Elemental Probabilities

In summary, for any risk assessment and any event we can use data from various local and external reputable sources available to us and interviews with key personnel; we

Fig. 10.3 Uniform
distribution f parameter x
between its estimated
extreme values Min, Max

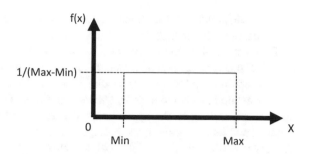

Fig. 10.4 A priori and a
posteriori distribution of a
parameter x between its
estimated extreme values 0.2
and 1

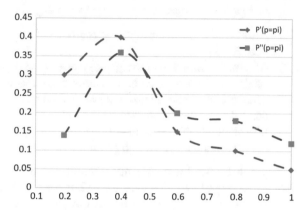

integrate these with any factual data the client may deliver to us; we adjust the values
to take into account the specific location, habits, and possible future conditions; and
finally, we obtain a framing range of probabilities and consequences for each type
of event and for each identified scenario.

> We live in a dynamic world of climate changes and environmental modifications. If any
> scenario is not present in an analysis it means that, at that time and with the data at hand,
> it was deemed as bordering on credibility, i.e., as having a probability of occurrence
> in the order of 10^{-6} to 10^{-5} (based on various industries' customary definition of
> credibility, as discussed earlier). We firmly discourage limiting risk assessments to
> "credible scenarios", as this common practice leads to bias and censoring the results[2]
> (Oboni and Oboni 2018). Events that are below credibility should fall out of the
> analysis "by themselves", and not because they are the object of arbitrary decisions.
>
> Bayesian updates can be developed, provided the a priori ranges of probabilities,
> consequences, or any parameter of interest are "wide enough".

Any risk assessment should be updated if the frequencies, intensities, patterns of
the hazards and/or vulnerabilities, or robustness of the systems change over time.

---

[2] *Because the credibility threshold may not be properly defined and people tend to instinctively place
it orders of magnitude higher than the technical definition.*

In particular, sudden and significant changes of climatic conditions should prompt a re-evaluation of the results of any risk assessment.

To conclude this section we need to talk again about the Bayesian inference model. The Bayesian inference model is indeed one of the cornerstones available to analysts among the tools for estimating probabilities and consequences. Allowing for rational updates when new data become available (from semi-static to real-time updates, depending on the applications and resources) will become the norm due to social and legal requirements, social and media pressure, etc. To allow Bayesian inference to be included (possibly at a later date of development of a risk management approach) a few conditions are required from day one:

(1) Always start by identifying hazards using threats-to and threats-from (see Sect. 8.2.1).
(2) All probabilities and consequences need to be expressed as ranges rather than single values, and uncertainty needs to be recorded. For example, for every parameter range there must be a justification recorded with the different opinions that led to its definition recorded as well.
(3) Inter-dependencies need to be transparently implemented, i.e., the methodology has to provide results for singe failures and domino effects.

Once these conditions are met, then on the basis of specific geotechnical (or other) data reports, weather data, etc., a Bayesian inference model can be developed to determine posterior probabilities. In time, as data are gathered and changes occur, probabilities will be seamlessly updated.

At each new data entry, at discrete time intervals, as required or in real time, depending on the application, it will be necessary to perform new probability-magnitude estimates for the hazards and their consequences. That will allows for Bayesian updates of the risks.

10.2   Probability of Failure in a Portfolio

10.2.1   Independent Elements

Let's use as an example a theoretical portfolio of independent dams. Each facility/dam has its own probability p_f of incurring Maximum Foreseeable Loss (MFL).

If the failure criteria states, very simply, that the failure of one single dam means the portfolio fails, if the MFL events are also independent, the overall system's probability of failure p_{fs} can be evaluated using the series probability Formula (4.1). This failure criteria would be interesting, for example, for an insurer that wants to decide whether to insure or deny insurance on a specific dam portfolio.

Independent Identical Elements
Let's first assume that the dams have the same p_f. The probability of failure of each dam is shown in Table 10.1 as $P = 1/T$ single dam. T would be the historic "frequency"

Table 10.1 Probability of failure of theoretical sample portfolios of identical dams

			Tailings dams World wide			
		Poor dams	Hydro dams world wide			
	Very poor dams				Best dams world-wide	
	Pf dam, identical for the three dams					
T	10	100	1000	10000	100000	1000000
P=1/T single dam	0.1	0.01	0.001	1.00E-4	1.00E-5	1.00E-6
Ptot(3)	0.271	0.0297	0.003	3.00E-4	3.00E-5	3.00E-6
Ptot(10)	0.651	0.0956	0.00996	1.00E-3	1.00E-4	1.00E-5
Ptot(100)	1	0.634	0.0952	9.95E-3	1.00E-3	1.00E-4

of failure of the selected type of dam. The coloured domains in Table 10.1 are based on statistics related to the world-wide dams portfolio we developed in (Oboni and Oboni 2013). For example, poor dams (orange colour) have p_f between, say, $1/10 = 0.1$ and $1/100 = 0.01 = 1.0^{-2}$. Excellent dams (best dams world-wide, blue colour) have p_f between, say, $1/100,000 = 1.0^{-5}$ and $1/1,000,000 = 1.0^{-6}$. The other "categories" lie in-between as indicated. Ptot (3), Ptot (10), and Ptot (100) indicate the p_{fs} for portfolios of 3, 10, 100 dams of identical $p_f = 1/T$ single dam.

The same information can be show in graphic form (logarithmic scales) in Fig. 10.5. Obviously, if the quality of the dams is poor to very poor the chance of a system failure is high, even for a small portfolio (3 dams).

However, as soon as the quality of the dams reaches world-wide standards for hydro-dams (not agricultural dams in the United States, for example, or decrepit structures anywhere in the world) the system's probability remains below $1/100 = 0.01 = 1.0^{-2}$ and for portfolios of, for example, 10 dams or fewer, it is below $1/1000 = 0.001 = 1.0^{-3}$.

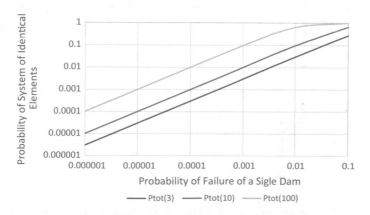

Fig. 10.5 Probability of failure of theoretical sample portfolios of identical dams

Independent Different Elements

Let's now look at different, independent dams. We recently developed a tailings dam system risk assessment for a client owning/operating 15 dams. The results are displayed in Fig. 10.6.

Each yellow vertical bar corresponds to a dam in the system. The bottom of the bar corresponds to the optimistic estimate of the probability, while the top corresponds to the pessimistic one, based on extant uncertainties. An explanation on how to derive these values is given in Chap. 11. As can be seen, the quality of the dams in the portfolio is extremely variable. The horizontal grey lines give the historic world-wide tailings dam performance (Table 4.1, used as bench-mark). The two green rectangular

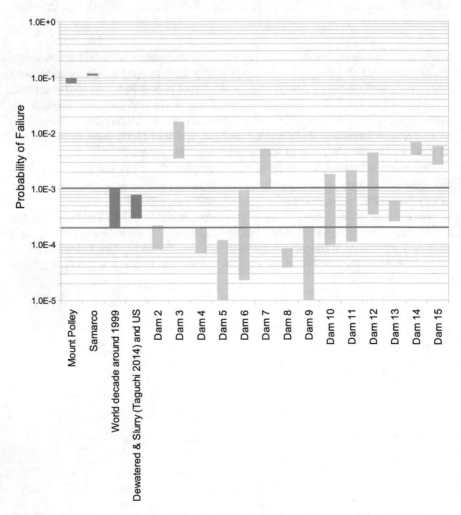

Fig. 10.6 Relative probabilities of failure and historic benchmark for a portfolio of 15 dams

spots at the top left correspond respectively to the p_f of Samarco and Mount Polley evaluated using the data that were available to us before their respective failures.

Using the client's data, we evaluate the system's probability p_{fs} on the maximum estimates of the client's dams (yellow bars) to be $4.37 * 10^{-2}$/year or roughly (4.5/100)/year. As it can be immediately seen, the system is heavily influenced by the few "poor dams" present in the system. This stresses the need to perform quantitative prioritization of systems, in order to swiftly detect the "bad apples" which affect the whole system.

As additional proof of the statement above, we added to the previous analysis first one and then both, green examples (top left) of Fig. 10.6. As stated above, those correspond to the estimates of Samarco and Mount Polley, developed with the data that were available before their respective failures. This addition corresponds to considering one or two extra "bad apple" in the portfolio. As expected, the system's probability increases to $1.39 * 10^{-1}$/year (13.9/100)/year if we include Mount Polley, and further to $23.4 * 10^{-1}$/year (23.4/100)/year if we include both Mount Polley and Samarco. This shows again the serious effect of "bad apples" on a whole system.

Conclusions on Independent Elements

In a real-life system of independent dams it is more efficient to define (and mitigate) the "bad apples" rather than evaluating "cumulative" potential losses to justify more reserves. This bears particular interest for insurers interested in insuring entire portfolios.

10.2.2 Dependent Elements

As introduced earlier, the case of dependent facilities is very complex. The dependencies can be physical, intrinsic or external, related to common cause failure (CCF), etc.

Dependent facilities also have the potential to generate MFLs which are larger than the single facility failure would generate by domino effects. Thus the reasoning has to cover simultaneously the probability and the consequences, i.e., a full risk assessment has to be performed.

Example of Two Cascading Dams

Figure 10.7, again from a real-life study, shows the result for one dam (Dam 1, blue line), compounded with the potential failure of a second one (Dam 2). The compounded case is the orange line.

As can be seen, the compounded failure of the two dams is less likely than the failure of Dam 1 alone, but the consequences are more than double (of course, depending on the reservoirs volumes and many other factors to be determined).

Conclusions on Interdependent Elements

In a real-life portfolio, dependent failures must be studied with attention. Oftentimes risks from dependent failures may be higher or smaller than what intuition suggests. There is no "one size fits all" intuitive solution to these estimations.

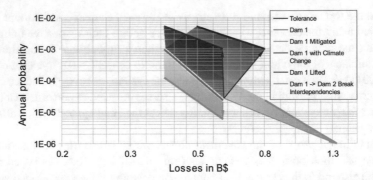

Fig. 10.7 Result for one dam (Dam 1), blue line, compounded with the potential failure of a second one (Dam 2). The compounded case is the orange line

10.2.3   Summary of Conclusions on Elements' Portfolios

(1) The probability of simultaneous failure of independent dams is the product of their probabilities, i.e., an extremely small value for "normal dams". For example three dams with $p_f = 0.001$ (1/1000) give 1/1,000,000,000 (one to one billion).

(2) However, should those dams be interdependent, the system's probability of failure could rise as high as the original $p_f = 0.001$ (1/1000), with consequences increasing due to domino effects.

(3) In a real-life system of independent dams it is more efficient to define (and mitigate) the "bad apples" rather than evaluating "cumulative" losses to justify more reserves. The dams are then to be evaluated jointly (simultaneous failure, i.e., product of the dams' probabilities) to determine the maximum portfolio potential loss and its probability.

(4) In a real-life portfolio dependent failures have to be studied with attention. Oftentimes risks from dependent failures may be higher or smaller than what intuition suggests. There is no "one size fits all" intuitive solution to these estimations.

References

Ang A H-S, Tang WH (1975) Probability concepts in Engineering Planning and Design, Vol. I, John Wiley and sons

Carlin, BP, Louis, TA (2008). Bayesian Methods for Data Analysis (Third ed.). CRC Press

NIST/SEMATECH e-Handbook of Statistical Methods, April, 2012

Oboni C, Oboni F (2013) Factual and Foreseeable Reliability of Tailings Dams and Nuclear Reactors -a Societal Acceptability Perspective, Tailings and Mine Waste 2013, Banff, AB, November 6 to 9, 2013

Oboni C, Oboni F (2018) Geoethical consensus building through independent risk assessments. Resources for Future Generations 2018 (RFG2018), Vancouver BC, June 16–21, 2018

Oboni F, Oboni C (2016) The Long Shadow of Human-Generated Geohazards: Risks and Crises, in: Geohazards Caused by Human Activity, ed. Arvin Farid, InTechOpen, ISBN 978-953-51-2802-1, Print ISBN 978-953-51-2801-4

[Oboni et al. 2014] Oboni F, Oboni C, Caldwell J (2014) Risk assessment of the long-term performance of closed tailings, Tailings and Mine Waste 2014, Keystone CO, USA, October 5–8, 2014

Rosenblueth E (1975) Point Estimates for Probability Moments. Proceedings of the National Academy of Sciences of the United States of America PNAS 72(10): 3812–3814

Straub D, Grêt-Regamey A (2006) A Bayesian probabilistic framework for avalanche modelling based on observations, Cold Regions Science and Technology 46: 192–203

Chapter 11
Dam Stability Failures

> In this chapter we focus on the very common stability failure mode to demonstrate how it can be evaluated

Failure analysis leading to the risk assessment of dams is a complex process. It requires modelling the complex relationships between hazards and the potential response of the dam system. Salmon and Hartford (1995) used Event Tree Analyses (ETAs). When using an ETA (Viseu and Betamio de Almeida 2009), the initiating event (e.g., hazard) can be dc-convoluted in a logical structure, so that all events, which can cause a failure or malfunctioning of the system, can be evaluated. Taguchi (2014) used several trees to evaluate the probability of failure (not the risk) of several types of TSFs.

In this book, as already stated, we will focus on a very common failure mode, i.e., instability, as an example, but similar approaches can be develop for any of the pertinent failure modes.

11.1 Classic Approach

11.1.1 Linking FoS to P_f: General Simplified Approach

In a classic stability analysis the Factor of Safety (FoS) is defined as the ratio between the capacity (C_S) of the slope to withstand failure under the demand (D_S, driving forces) (Eq. 11.1) generated by gravity, water table and, in the case of pseudo-static analysis, a horizontal acceleration due to a quake.

$$FoS = C_S/D_S$$

$$(11.1)$$

© Springer Nature Switzerland AG 2020

F. Oboni and C. Oboni, *Tailings Dam Management for the Twenty-First Century*,
https://doi.org/10.1007/978-3-030-19447-5_11

The analyses can be performed in drained conditions (effective stress analyses, or ESA) or undrained conditions (undrained stress analyses, or USA) based on a number of considerations.

As neither C_S nor D_S are perfectly known due to the numerous uncertainties surrounding the geomechanical properties, hydrological conditions, etc., both values should be assumed to be stochastic variables. Thus it is spontaneous to express the probability of failure p_f of a structure (a slope) (Eq. 11.2) as

$$p_f = p(FoS \leq 1) \text{ or alternatively } p_f = p(C_S \leq D_S) \tag{11.2}$$

The limit conditions FoS = 1 or $C_S = D_S$ denote meta-stable equilibrium (like the toss of a coin), with a theoretical p_f value at 50%.

At this point we have a benchmark range $(2 \times 10^{-4}$ to $10^{-3})$ (Oboni and Oboni 2013), the credibility threshold at, say 10^{-6} (see Sect. 7.3) and the meta-stable p_f value at 50% as described above. The goal of the approach will be to see where each structure in the portfolio lies with respect to these framing values.

11.1.2 Linking FoS to P_f: Three Simplified Methods

We start with **Method 1**, transforming a "simplified" deterministic slope stability analysis into a "probabilistic one" using the $p_f = p(FoS \leq 1)$ simplification (Eq. 11.2). This is like considering the FoS a stochastic variable, for which the members of the assessing team define the expected value, min, max (either subjectively or from various analyses, or both) and then p_f (stability) is calculated as $p(FoS \leq 1)$. Note that values of p_f (stability) can be estimated using other modelling procedures:

Method 2: Use point estimates methods (like that of Rosenblueth 1975) and various later extensions) to define the variability of the FoS and then pf (stability) is calculated again as $p(FoS \leq 1)$

Method 3: Use probabilistic stability analysis methods as in (Oboni and Bourdeau 1983).

After defining the p_f (stability), which should include uncertainties of the material, construction care and design, either ETAs or FTAs can be constructed to account for monitoring and maintenance, leading to "the complete" p_f (see Taguchi 2014 for examples).

The p_f (stability) is the starting point of the ETA. Monitoring, repairs, etc. come in as branches leading to a possible reduction of the initial probability following procedure Methods 1, 2 or 3. It can be concluded that:

- Methods 1 and 2 plus ETA/FTA provide a good estimate of the p_f.
- Method 3 plus ETA/FTA is even better, but is generally reserved for very critical cases because it requires re-doing commonly available slope stability analyses performed by the project's engineers.

Table 11.1 An example of FoS obtained by engineers studying a dam

Section	FoS$_{Local}$	FoS$_{mid}$	FoS$_{Global}$
1—South	>1.5	1.24	1.36
2—North	>1.5	1.27	1.20
3—North	>1.5	1.26	1.26
4—South	1.02	>1.5	1.50
5—North	1.01	1.11	1.82

We focus on the North section of a dam where reportedly (see Tables 11.1 and 11.2):

- slopes are steeper than designed;
- failure is incipient toward the eastern abutment;
- there is a monitoring system, but there is limited information about the readings, frequency, etc.;

Table 11.2 Factor of safety details for two sections (North and South)

Section North	FoS	Section South	FoS
B–B′	1.302	–	–
A–A′	1.295	C–C′	1.448
646,290	1.265	646,290	1.645
646,356	1.02	646,356	1.573
646,406	1.245	646,406	1.336
646,456	1.279	646,456	1.578
646,506	1.264	646,506	1.395
646,556	1.132	646,556	1.313
646,630	1.109	646,606	1.451
646,656	1.246	646,656	1.08
646,706	1.079	646,706	1.441
646,756	1.025	646,756	1.423
646,806	1.017	646,806	1.312
646,856	0.998	646,856	1.39
646,906	1.192	646,906	1.502
646,956	1.202	646,956	1.356
647,006	1.117	647,006	1.358
647,056	1.059	647,056	1.394
647,106	1.004	647,106	1.583
647,156	1.235	D–D′	1.578
647,206	1.277	E–E′	1.393
647,256	1.35	–	–

- there is uncontrolled erosion of the downstream slopes.

Consequences: road failure, personnel safety hazards, mining infrastructure damage.

Table 11.1 shows a summary of the FoS obtained by the engineers.

Let's now apply Method 1 to the results of two cross-sections displayed in Table 11.2.

If we look at the first column F.S. (FoS) it is easy to evaluate "experimental values" of the FoS population as follows:

minimum = 0.998;
maximum = 1.350;
average = 1.168;
standard deviation = 0.115.

This leads to a coefficient of variation (COV) = standard deviation/average (%) of roughly 10%. Let's define the "tails" of the distribution of FoS as maximum-average and average-minimum and express them in number of standard deviations (sigma bounds).

As the "tails" are respectively only 1.48 sigma bounds and 1.58 sigma bounds the distribution would be a bathtub-shaped one, i.e., a U-shaped distribution. That may mean that we are sampling two different "families" of slope in that first column. It seems this corresponds to a geological/geotechnical/physical reality as at least one area of the North slope is less stable than the rest.

Let's suppose now, for the sake of the discussion, that we would come to the conclusion the last statement is not correct. Then we would conclude that the analyses have actually not uncovered the "true" max, min values of the FoS, and could "play around" with those extreme values to understand what happens to the distribution of FoS.

We would notice that using an empirical Beta distribution requires us to lower the lower bound (min) to 0.85 to get a pseudo-triangular distribution, which would lead to p_f (stability) $= 0.06 = 6 \times 10^{-2}$ (approx. 6%). Raising the upper bound at 1.55 would lead $p_f = 0.13 = 1.3 \times 10^{-1}$ (approx. 13%) and a pseudo-bell-shaped distribution. At this point we could build an ETA to complete the analysis of the case.

So, at the end of the day, by using a blended approach we can swiftly frame the problem at hand and allow a rational discussion related to the "homogeneity" of the considered slope.

Also, the 6–13% range evaluated with Method 1 sits well within our previous experience on similar cases. In Chap. 15, and particularly in Sects. 15.5.4 and 15.5.5 we will see that this theoretical range may even be a low estimate as it does not include defects and uncertainties that became known in the course of the study. That being said, any dam with a probability of failure above 1% (0.01) can be considered a very poor structure (see Table 10.1) and a candidate to failure. Incidentally, Mount Polley and Fundao Dam, evaluated with the data available before their respective failures had probabilities between 8 and 17% (see Sect. 11.2.2).

If we were to add the effects of monitoring and maintenance (unfortunately almost non-existent here) by using an ETA the final probability range would not change much from the one obtained with Method 1, in this particular case.

11.2 Semi-empirical Approach

In order to work within the effort limits normally consented for this type of risk assessment, especially if swift prioritization is the goal, we will now examine if a relatively recent set of empirical relations between FoS and p_f can be used. The approach was originally described in (Silva et al. 2008) (we call this the Silva-Lambe-Marr (SLM) method, after the authors), and later modified in (Altarejos-García et al. 2015).

As with all empirical and simplified approaches caution must be exerted before using these relations for tailings dams, in particular because neither publication makes either explicit mention of construction mode (downstream, centre-line, upstream) or provides a specific discussion of seismic loading, static and/or dynamic liquefaction. In our practice, we have modified and re-calibrated these approaches specifically for tailings dams.

The empirical relationships link the FoS (stability) defined by the geotechnical engineers in charge of a project, using classical stability methods, to the annual probability of failure.

11.2.1 Category of a Structure

In order to link FoS to p_f, the first step is to define the "category" of the structure under examination. This is done by examining sequentially the aspects of the design (D1, investigation; D2, testing; D3, analyses and documentation) and construction (CO) as well as operations and monitoring (OM). We will see these aspects appear in Table 11.4.

The methodology considers four categories ranging from I (best) to IV (poor, i.e., non-engineered) and experience has shown (over a rather large array of cases studied by the authors) that structures with high failure consequences are generally designed built and operated in such a way that they fall into category I. Of course, if a structure has received little or no engineering it will fall in category IV.

Figure 11.1 shows the relationship between FoS and the annual probability for the four categories.

If we consider, for example, a relative common practice choice of FoS = 1.3, we pull from Fig. 11.1 the values of the probability of failure displayed in Table 11.3 for the four categories.

The values shown in Table 11.3 mean that, based on the assumed common practice FoS = 1.3 and the historical values described earlier (the bench-marking range), the

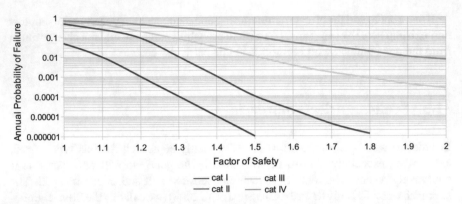

Fig. 11.1 pf-FoS empirical relationships per structure category as per (Altarejos et al. 2015)

Table 11.3 Variation of the probability of failure for constant FoS, but varying Category of structure

Category	FoS	Annual p_f
I	1.3	9.31×10^{-5} or appx 10^{-4}
II		5.93×10^{-3} or appx 6×10^{-3}
III		6.4×10^{-2} or 6.5%
IV		2.36×10^{-1} or 23.6%

world-wide portfolio of active dams is probably constituted by structures between Category I and II, and that those which have failed having likely belonged to Category II to IV.

It is interesting to look at the values of p_f for FoS = 1 in Fig. 11.1. The meta-stable condition leads to $p_f = 0.5$ for Category II, and slightly higher values for Cat. III and IV. We wonder what the reasoning of Silva-Lambe-Marr was, but notice some "observation points" that leave us quite surprised. In any case, once a structure has reached $p_f = 0.5$, all discussions become useless, as the structure has virtually failed. The other interesting point is Category I, with $p_f = 0.05$ for FoS = 1. In order to understand this apparently odd value, let's use an example from aerospace: if design, control, construction and monitoring are developed to the highest standards, as in aerospace, then a FoS near 1 can be accepted because the very narrow uncertainties lead to a lower probability of failure than in "normal" applications. Many rockets/missiles are built that way to maximize payload, and the media report corresponding failures at a rate that leaves the public baffled, but is not, after all, very significant considering the low-margin design.

Finally, if we look at the other end of the spectrum, we are not keen on using empirical relationship that go below credibility. The nuclear industry has shown, for example, that extremely low theoretical probabilities result in significantly larger actual real-life probabilities and rates of failure (Oboni and Oboni 2013). Likewise, we "do not trust" very large factors of safety (above 2), because the models used are generally incomplete and approximate.

Now that the relations have been discussed and somewhat framed in reality, Table 11.4 is the key to determine the category of a structure based on a number of its characteristics. To display it in this book, we have adapted it from the original versions of SLM and (Altarejos et al. 2015). The original papers give detailed explanations on how to determine the category of a structure and suggests interpolation techniques to cover intermediate cases. That is necessary because a structure has generally varying degrees of quality over the various areas of characterization. In our dam-specific version we look at thirty diagnostic points, including design, construction, liquefaction potential, internal erosion potential, management, monitoring and maintenance as well as other specific dam-related points, and adopt a min-max approach to define a range of probabilities of failure to avoid conveying a sense of false "accuracy". As Fig. 10.5 makes clear, the level of uncertainty can vary quite considerably from dam to dam.

Table 11.4 How to determine a geo-structure category

Category	Design			Build	Operate
	D1	D2	D3	CO	OM
	Investigation	Testing	Analyses and documentation	Construction	Operations and monitoring
I Best	Evaluate design and performance of nearby structures	Run lab tests on undisturbed specimen at field conditions	Determine FS using effective stress parameters based on measured data for site	Full time supervision by qualified engineer	Complete performance program including comparisons between predicted and measured
	Analyze historic aerial photos	Run strength tests along field effective total and effective stress paths	Consider field stress path in stability determination	Construction control tests by qualified engineers and technicians	no malfunctions (cracks slides etc.)
	Locate all non uniformities	Run index field tests (field vane, come penetrometer) to detect anomalies	Prepare flow nets for instrumented sections	No errors or omissions	Continuous maintenance

(continued)

Table 11.4   (continued)

Category	Design			Build	Operate
	D1	D2	D3	CO	OM
	Investigation	Testing	Analyses and documentation	Construction	Operations and monitoring
	Determine site geologic history	Calibrate equipment and sensors prior to tests	Predict pore pressure and other relevant performance parameters for instrumented sections	Construction report clearly documents construction activities	
	Determine subsoil profiles using continuous sampling		Have design report clearly document parameters and analyses used for design		
	Obtain undisturbed samples for testing		No errors or omissions		
	Determine field pore pressure		Peer review		
II Above Average	Evaluate design and performances of nearby structures	Run standard tests on undisturbed specimen	Determine FS using effective stress parameters and pore pressure	Part time supervision by qualified engineer	Periodic inspection by qualified engineer
	Exploration program tailored to project conditions by qualified engineer	Measure pore pressure in strength tests	Adjust for significant differences between field stress paths and stress path implied in analysis that could affect design	No errors or omissions	No uncorrected malfunctions
		Evaluate differences between laboratory tests conditions and field conditions			Selected field measurements
					Routine maintenance

(continued)

Table 11.4 (continued)

Category	Design			Build	Operate
	D1	D2	D3	CO	OM
	Investigation	Testing	Analyses and documentation	Construction	Operations and monitoring
III Average	Evaluate performances of nearby structures	Index tests on samples from site	Rational analyses using parameters inferred from index tests	Informal construction supervision	Annual inspection by qualified engineer
	Estimate subsoil profile from existing data and borings				No filed measurements
					Maintenance limited to emergency repairs
IV Poor	No field investigation	No laboratory tests on samples obtained at the site	Approximate analyses using assumed parameters	No construction supervision by qualified engineer	Occasional inspection
				No construction control tests	No field measurements

Adapted from SLM and (Altarejos et al. 2015)

11.2.2   Some Advisable Limits

By using the empirical relationships described above, data that were known before the accidents, we evaluated the following ranges of probabilities for two recent failures, which we used as a further bench-marking element in the portfolio risk assessment (Fig. 10.5):

- **Mount Polley**: Operations were going to add a buttress at the toe, a bulldozer and people were working at the toe. Nothing was noted until short before the failure. Estimated $p_f = 0.08 - 0.1$.
- **Fundao Dam**: Plans to build a buttress were laid out. Leaks were noted before failure. Estimated $p_f = 0.11 - 0.17$.

Again, it is strongly recommended to comply with the "credibility threshold" (see Sect. 7.3) and therefore not to state probabilities of failure lower than 10^{-5} to 10^{-6}.

Consider as well what very large FoS may mean and avoid blindly accepting values of the FoS > 2, as they may be meaningless.

11.2.3   Does It Really Work?

Here is an example of "simplified" transformation of a deterministic slope stability
analysis into a "probabilistic one" using the simplified approach described in Silva
et al. (2008). The dam under consideration is well maintained and not damaged.
However, there was little control during construction and only partial monitoring
and observations. So, SLM method would propose a Class II–III structure, which is
definitely not a high quality dam.

With a FoS between 1.09 and 1.36 obtained by the engineers through their classic
slope stability analyses, we would have a $p_f = 10^{-1}$ to 5×10^{-2} or 5×10^{-2} to 10^{-3}.

It is now easy to include progressive release of attention, analysis of anomalies such as
deformations ... and evaluate the resulting increased p_f. It would then also be easy to
simulate progressive increase of attention, maintenance, monitoring, or hardening of
the structure through repairs and better monitoring, maintenance and see the effects.

We have applied this type of "symptom-based" approach, similar to the SLM
approach, to a variety of cases around the world, ranging from the evaluation of
unexploded ordnance risks in countries such as Lao PDR and Cambodia to the risk
assessment of mountainous roads networks. The success of the approach has been
measured by third parties through extensive reality checks.

For dams, we include thirty diagnostic points specific to tailings dams (see
Sect. 11.2.1). The range of probabilities stems out of a "optimistic" and a "pessimistic" evaluation of the thirty diagnostic points (a re-calibration of Table 11.4).
As many forensic studies have shown, the failure of a dam can rarely be attributed
to a single cause, but rather to a set of conditions (hence the thirty diagnostic points)
which, together, can bring the category evaluation to a higher value and thus significantly affect the probability of failure even if the FoS seems "reasonable".

In this type of application where the knowledge about symptoms and dysfunctions
is encoded to finally provide a probability of an event, the "ignorance"—i.e., "not
knowing enough about a symptom"—is an extremely important piece of information
and is carefully used in the analyses. This applies, for example to boreholes and
instrumentation that may be concentrated (for practical/access reasons) in an area
and thus generate a lack of information that is geographically well distributed.

Of course, as new information becomes available, the category of the structures
will change (hopefully towards a higher category) and Bayesian updates can be performed. The benchmarking against the world portfolio and recent failures is preserved
to help understating emerging trends.

Before closing this review of semi-empirical methodologies three more points
must be discussed:

(1)   Is there a correspondence between the results of a semi-empirical method and
      a direct slope stability approach?
(2)   Can a semi-empirical approach be used for pseudo-static analyses as well?
(3)   What about liquefaction?

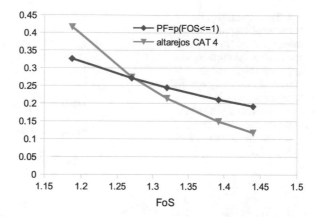

Fig. 11.2 Comparison of the direct geomechanical probabilistic approach (red) and the semi-empirical method results for cohesive soils

These three points are discussed below.

We have carried out a number of trials using a probabilistic geomechanical approach and compared the results to the empirical methodology. These were based on cohesive and granular slopes of non-engineered materials (natural soils). In each case found comparable results (that is, within the same order of magnitude) of the probability of failure, as depicted in Figs. 11.2 and 11.3, for category IV, i.e., non-engineered structures. NB: the probability of failure of a real-life dam lies in a range of six orders of magnitude. World-wide historic tailings dams performance lies within one order of magnitude. Finding a correspondence between a theoretical and empirical pf within one order of magnitude is sufficient for the purposes of portfolio prioritization, given all the other present uncertainties. Thus it is assumed that the direct geomechanical probabilistic approach and the empirical method deliver comparable results for non-engineered (natural) materials.

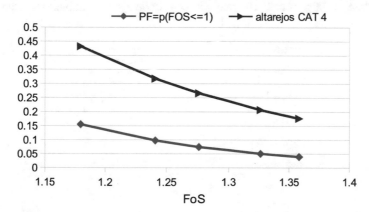

Fig. 11.3 Comparison of the direct geomechanical probabilistic approach (red) and the semi-empirical method results for granular soils

Fig. 11.4 Comparison of
the annualized probability of
failure derived from a direct
geomechanical probabilistic
approach (blue) and the
empirical method (green) for
various return periods
(horizontal acceleration) in
cohesive materials. The
slopes are the same as those
in Fig. 11.2

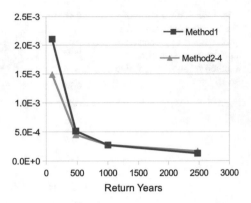

Engineered materials have fewer uncertainties than natural ones, as they are selected and carefully compacted under controlled water content. This reduces their variability. Thus we consider that categories III, II, and I of the empirical method reflect the reduction of uncertainties leading to lower probabilities for similar theoretical FoS. Controls and monitoring further amplify these effects.

As per the evaluation of pseudo-static conditions, our tests have also shown good agreement between the two methodologies. As Figs. 11.4 and 11.5 show, for cohesive, respectively granular materials, the annualized probability of failure can be higher for low magnitude, more frequent events than for high magnitude, long-return events. NB: the reason for this apparently odd statement is very simple. The pf is equal to the pf under horizontal acceleration multiplied by the annual probability of the quake generating that acceleration. Low magnitude events are more frequent, so the product can be higher (it depends, of course on the FoS-acceleration relationship.

So, what the graphs of Figs. 11.4 and 11.5 show is that if one looks at the annualized probability of failure, the smaller events may give a higher annual probability of failure than the larger ones! The implications for design are simple: one must check

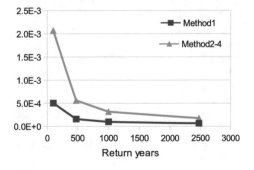

Fig. 11.5 Comparison of the annualized probability of failure derived from a direct geomechanical probabilistic approach (blue) and the semi-empirical method (green) for various return periods (horizontal acceleration) in cohesive materials. The slopes are the same as those in Fig. 11.3

the dam slope for high frequency, less intense events as well. That may be a way to avoid what happened to many dams that actually failed under "weak" earthquakes.

11.2.4   A Recent Example

Let's look at a recent dam failure example, namely the Cadia Ridgeway Mine in Australia already referenced in Sect. 9.2.1.

What We Do and Do Not Know about the Tailings Dam Failure
We have not had access to any information about this dam except what was available in the media before the Expert review, but it is not important for this discussion as we want to discuss tailings dams' seismic probability of failure. We do not know how the engineers evaluated the dams at this mine from a seismic point of view. We do know (because it is public knowledge) that the mine temporarily closed last year after a magnitude 4.3 earthquake hit. Furthermore, a couple seismic events reportedly occurred in the region prior to the recent failure.

We also know that forensic investigators/independent expert panels oftentimes invoke the possibility of "minor quakes" as triggers of tailings dams failures. For example for the Samarco tailings dam, the panel found that three small seismic shocks "triggered" the collapse. The panel also indicated the dam's failure was reportedly "already well advanced", caused by a string of design and maintenance failures.

Common practice for seismic pseudo-static analyses for tailings dams would be to:

- follow jurisdictional codes;
- look at the Maximum Credible Earthquake (MCE) or some code imposed level;
- select the corresponding Peak Ground Acceleration (PGA);
- perform pseudo-static stability analyses with that PGA (or a fraction of it, as defined by codes);
- ensure the dam attains a code-imposed FoS under those conditions.

Let's note the minimum FoS under those conditions may be as low as 1.1 following some national codes.

Tailings Dams Risk Management Considerations
The approach described above is absolutely common and unhesitatingly trusted by basically all engineers and regulators we know. Let's take a big risk here (pun intended) and question that credo using a hypothetical dam.

Assume for a minute that the hypothetical dam is designed for a 1/2500 earth-quake, with a corresponding PGA of 0.075 (g) and the selected 50% PGA value as horizontal acceleration. (NB: The 50% "rebate" is allotted in some jurisdictions because the foundation soils are of good quality without really wondering how local-ized geological variations may influence that quality.) For the sake of this discussion we have calculated the dam's downstream slope FoS for the static case, ensuring it was code compliant; a 1/2500 earthquake ensuring again it was code compliant; and finally also a 1/100 earthquake, which, by the way has only a PGA of 0.012 (g). The codes do not ask for the probability of failure and do not impose a FoS value for this type of event.

With a code-compliant dam we proceeded to calculate the probability of failure for the three cases above (p_{fstat}, $p_{f1/2500}$, $p_{f1/100}$). The probability that the FoS will be smaller than 1 (see Sect. 11.1) is used as the probability of failure. This is a simplification, as the mathematically correct formulation is a bit more complicated, as noted in the referenced section.

We Used Two Different Approaches for the Evaluation
Now comes the big point of this discussion. Actually, the static case occurs "every year", right? However, in the first approximation the 1/2500 does occur "only" 1/2500. Please forgive us for this gross approximation and remember that two 1/2500 quakes are always possible in the same year. This is also true for the 1/100 quake, and so forth. What that means is that the probability of failure under seismic event is actually a conditional probability equal, for example, respectively to $p_{f1/2500}$ divided by 2500 and $p_{f1/100}$ divided by 100. It is possible that the yearly probability of failure for a minor quake be higher than for a stronger phenomenon.

Thus, after reading the Expert review report we can summarize our interpretation of Cadia tailings facility failure (http://www.newcrest.com.au/media/market_releases/2019/Report_on_NTSF_Embankment_Slump.pdf) and risk considerations as follows:

- It is time to stop using misleading factors of safety to characterize dams "safety." FoS has the unfortunate tendency to foster excessive audacity, especially under seismic conditions of various intensity.
- Sometimes there are simple explanations to apparently complex problems, which make it possible to avoid invoking "complexity" and clarify issues.
- Those simple explanations lead to reasonable prioritization of dams portfolios and to sustainable, ethical mitigative roadmaps.

References

Altarejos-García L, Silva-Tulla F, Escuder-Bueno I, Morales-Torres A (2015) Practical risk assessment for embankments, dams, and slopes, Risk and Reliability in Geotechnical Engineering, ed. Kok-Kwang Phoon & Jianye Ching, Chap. 11, Taylor & Francis

Oboni C, Oboni F (2013) Factual and Foreseeable Reliability of Tailings Dams and Nuclear Reactors -a Societal Acceptability Perspective, Tailings and Mine Waste 2013, Banff, AB, November 6 to 9, 2013

Oboni F, Bourdeau PL (1983) Determination of the Critical Slip Surface in Stability Problems - Proc. of IVth Int. Conf. on Application of Statistics and Probability in Soil and Structural Engineering, Florence. Università di Firenze (Italy) 1983, Pitagora Editrice, pp 1413–1424

Rosenblueth E (1975) Point Estimates for Probability Moments. Proceedings of the National Academy of Sciences of the United States of America PNAS 72(10): 3812–3814

Salmon G, Hartford D (1995) Risk analysis for dam safety. Part I of II. Int. Water Power & Dam Construction 46(3): 42–47

Silva F, Lambe TW, Marr WA (2008) Probability and Risk of Slope Failure, Journal of Geotechnical and Geoenvironmental Engineering 134(12) https://doi.org/10.1061/(ASCE)1090-0241(2008)134:12(1691).

Taguchi G (2014) Fault Tree Analysis of Slurry and Dewatered Tailings Management – A Frame Work, Master's thesis, University of British of Columbia

Viseu T, Betamio de Almeida A (2009) Dam-break risk management and hazard mitigation, in: Dam-break Problems, Solutions and Case Studies, eds. D. De Wrachien, S. Mambretti, pp. 211–239. WIT Press

Chapter 12
Consequences

This chapter explains how to consider the multidimensional aspect of hazard consequences.

The consequences to be studied are generally perceived as mainly dependent on the target readership of the risk assessment: insurance companies, interested in some specific loss type; owners wanting to perform RIDM on mitigation; politicians looking to understand comparative risks of various energy sources; authorities wanting to assess societal risk and cover public protection issues (Darbre 1998).

The Australian Landslide Practice Note Working Group (AGSLT 2007) states, for example, that the elements "at risk" will include:

- Property, which may be subdivided into portions relative to the hazard being considered.
- People, who either live, work, or may spend some time in the area affected by landsliding.
- Services, such as water supply or drainage or electricity supply.
- Roads and communication facilities.
- Vehicles on roads, subdivided into categories (cars, trucks, buses). These should be assessed and listed for each landslide hazard.

For some cases, other consequences may also have to be considered. For example:

- Environmental, where the elements at risk are environmental (rather than man-made), such as forests or water bodies.
- Social, where the consequences of the landslide may have an impact on social conditions, such as the cost of disruption to traffic where roads are affected.
- Political, where the consequences may not be acceptable in political terms.

© Springer Nature Switzerland AG 2020
F. Oboni and C. Oboni, *Tailings Dam Management for the Twenty-First Century*,
https://doi.org/10.1007/978-3-030-19447-5_12

NB: we have edited the above text to eliminate a misleading confusion between risks and consequences. We have quoted the term "at risk" for the same reasons.

As stated earlier, we contend this siloed vision of risk assessments. The hazard and risk register (i.e., the database that contains all the information on the elements, impinging hazards, probabilities consequences) should be maintained silo-free and general. This will make it possible to drill out information through the life of the system without ever wasting any information.

12.1 Dimensions of Failures

Each risk assessment requires the definition of consequences for the various scenarios considered for each specific element of the system under consideration. As stated in Sect. 7.6, consequences are multidimensional and the final result is the sum of the various dimensional components. It is a serious mistake to "select" the worst dimension as the value for the consequence, as it is oftentimes suggested in risk assessments instructions to prepare FMEA, risk matrix approaches.

Here is a list of four dimensions identified for various real-life studies with their (also additive) sub-components:

Direct physical losses:

- Fixed equipment
- Mobile equipment
- Infrastructures
- Business interruption

Health and Safety: (none of these include chronic effects—see environmental)

- Harm to workers
- Harm to outsiders

Environmental: (if dust and other by-products, for example contaminated water, are released at code limit values we generally assume there are no chronic effects)

- Water releases
- Tailings releases
- Hazardous materials releases
- Dust releases

Reputational Damages. These will arise for example from:

- chronic releases—for example in high-density residential areas—of materials which are perceived as hazardous, even if they are code compliant;
- incidents and accidents that destabilize the public;

- alterations of the perceived environment (for example, change of colours), even if they bear no measurable consequences;
- lack of communication;
- lack of empathy and care.

Reputational damages may lead to crises of various types and depth (cost), legal proceedings and consequences, and ultimately suspensions of licenses, business interruption, etc. The areas of the environment reportedly more prone to generating these damages are:

- highways,
- residential quarters,
- water courses receiving mine outflows (including accidental ones).

Table 12.1 depicts an example from a real-life study, not a general rule. It defines the potential consequences evolution from minor to ultimate for each consequence dimension.

12.2 Dam Breach Consequences

As Table 12.1 shows, various mechanisms will lead to the possible ultimate failure, i.e., a dam breach. The consequence assessment is therefore a critical component of carrying out a TSF risk assessment.

Unfortunately, there is a great deal of uncertainty in assessing a TSF failure and it is not possible to produce a definitive deterministic analysis. Consequently these studies oftentimes include a sensitivity analysis of hypothetical failures that would produce a range of possible consequences that can be used in a risk assessment. However, that is not sufficient.

A preliminary analysis of potential dam breach mechanisms may conclude, for example, that the following three failure mechanisms form the basis for assessment of a tailings outflow model study:

- Flood-induced over-topping (rainy day scenario);
- Slope-instability-induced over-topping (sunny day scenario);
- Piping failure (sunny day scenario).

In addition to the three failure mechanisms the uncertainty related to tailings rheology and their flow properties should be incorporated in the outflow model. Three scenarios for each of the three failure modes can then be assumed, for example, with the following characteristics:

- Tailings flow is a Newtonian fluid (water);
- Tailings flow is a Bingham fluid with average rheological properties;
- Tailing is a Bingham fluid with a high concentration of solids.

Table 12.1 Example from a real-life study, not a general rule

General element of the system	Hazard scenario	Vector, threat-from code	Potential consequences evolution from minor to ultimate			
			Direct physical losses	Health and safety	Environmental	Reputational damages
Pipelines [tailings and dewatering (pond, polishing ponds, ditches)]	Traffic (includes contractors and snow removal collisions)	G4, G5, P1	From very limited (if caught in time) to major dam damages and finally dam break (interdependent failure of dam)	From very limited to harming workers (dam break consequences studied in dams)	From very limited (if caught in time) to major environmental damages and finally dam break (interdependent failure of dam)	Generally limited unless failure evolves in dam break
	Freezing (includes deformations at flanges)	G4, G5				
	Sabotage/vandalism, cyber	P4, A1, A2, A3				
	Corrosion and intrinsic mechanical failures	F1				
	Human (communication between mill and tailings operation), IT	P1, E1, F1				
Confining structures (and their content)	Geotechnical/geological and climatological, hydrological	G1, G6	As above with possible "sudden failures". Additionally inadequate storage capacity may lead to BI, costly system alteration requirements	From very limited to harming workers and possibly general public downstream	From very limited to major environmental consequences outside of the perimeter	We consider that any dam issue would be covered by local and provincial media and further, with great reputational damages
	Construction/building operations	P1, P3, P4				
	Operation, maintenance and monitoring during service and record keeping, IT	P1, P2, P3, E1				
	Sabotage/vandalism, cyber	A2, A3				

(continued)

Table 12.1 (continued)

General element of the system	Hazard scenario	Vector, threat-from code	Potential consequences evolution from minor to ultimate			
			Direct physical losses	Health and safety	Environmental	Reputational damages
Outlet structures (spillways/pond to pond culverts)	Geotechnical/geological and hydrological (capacity, durability)	G1, G4, G5, G6	From very limited (if caught in time) to major dam damages and finally dam break (interdependent failure of dam)	From very limited to harming workers (dam break consequences studied in dams)	From very limited (if caught in time) to major environmental damages and finally dam break (interdependent failure of dam)	Generally limited unless failure evolves in dam break
	Operations (service, emergency access), IT	E1, E2, P1, P3				
	Water balance management	G2, G6				
	Climate (includes extreme freezing)	G5				
	Sabotage/vandalism, cyber	P4, A1, A2, A3				

Thus the evaluation of a dam breach outflow requires attentive analyses using complex rheological relationships, such as Bingham fluids. The parameters of the Bingham fluid equation are not easy to determine.

For each failure mechanism above, a set of dam breach parameters can be combined with the three alternative flow characteristics to produce a representative set of dam breach scenario combinations for assessment in a dam break study. The scenarios matrix then incorporates the range of likely outflow volumes from "smaller" to "larger" breach events. Although this may not be considered an exhaustive set of the possible outcomes, it may be assumed to provide a basis for screening the range of possible outflow extents.

The storage volume for the dam during normal operations includes tailings, supernatant pond, and freeboard below the design elevation crest. The storage volume curve is generally available from the design engineers. The storage curve provides the potential total volume for the dam breach scenarios. This is then modified to reflect the estimated outflow from the scenarios where the outflow volume is only a fraction of the total potential volume.

12.2.1   Tailings Rheology

The tailings rheology can of course play an important role in the fluid dynamics and spatial extent of the dam breach flood. The tailings rheology in the TSF can vary temporarily and spatially (horizontally and vertically) depending on numerous factors such as solid concentration, mineralogical composition, tailings delivery system, local hydrology and hydro-geology, and operating protocol. When the concentration of solids increases, the tailings rheology starts deviating from Newtonian fluid behaviour to become a visco-plastic fluid. Furthermore, tailings can liquefy under a critical shear stress value (the yield stress or yield strength). Bingham fluids are a subset of visco-plastic fluids that are well-suited for describing "muddy flows" where there is a high concentration of fine-grain sediment without the presence of much sand or coarser particles.

We have seen cases where efforts were made to divide tailings volumes into two or more classes as a function of volumetric concentrations (for example, normal and high or somewhat consolidated). Volumetric concentration C_v is the percent volume of solids to the total mixed volume of solids and water (%v/v).

Examples of the histograms for this type of data are shown in Figs. 12.1 and 12.2. In this example the clay mineral content for the tailings was assumed to be minimal based on extant description of the tailings. The particle density for the tailings solids in the TSF was 2.8 t/m^3, based on reported particle densities. These graphs can be used to determine the rheological parameters for the Bingham fluid.

The primary Bingham fluid parameters are:

- Yield strength (τ_y) in Pa;
- Dynamic viscosity (μ_m) in Pa s;

Fig. 12.1 Example of C_v (%v/v) histogram from a tailings pond

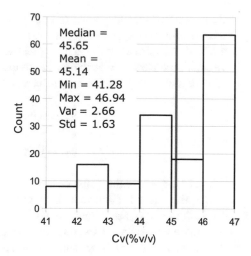

Median =
45.65
Mean =
45.14
Min = 41.28
Max = 46.94
Var = 2.66
Std = 1.63

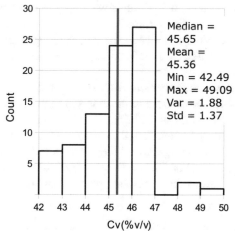

Median =
45.65
Mean =
45.36
Min = 42.49
Max = 49.09
Var = 1.88
Std = 1.37

Fig. 12.2 Example of C_v (%v/v) histogram from a tailings pond. It is interesting to compare this graph with the one in Fig. 12.1. They display strong differences in the high values and difficult to interpret distributions

- Fluid density ρ_m in kg/m³.

The fluid density is the total density of the fluid mixture, i.e., sediments and water. Tailings may be assumed to be fully saturated.

Previous work on Bingham fluids have found a correlation between these parameters and the concentration of solids, particle size and presence of clay minerals where τ_y exponentially increases as a function of C_v, τ_y will change slowly with increasing V_v (Julien 2010). At some point this changes such that small changes in C_v can result in significant changes in τ_y.

Julien (2010) provides a set of empirical equations to estimate τ_y and μ_m as a function of material type and volumetric concentration. In the example of Figs. 12.1 and 12.2 the empirical equation for typical soils where the presence of clay minerals is minimal, the particle distribution is composed of very fine sand, silt, and a minor portion of clay-size particles were selected. The equations are:

$$\tau_y = 0.005 \times 10^{7.5Cv} \tag{12.1}$$

$$\mu_m = 0.001 \times 10^{8.0Cv} \tag{12.2}$$

Figure 12.3 shows the graph of τ_y and μ_m as a function of C_v.

Using the equations above and the graph, it is easy to determine the following Table 12.2 τ_y and μ_m as a function of C_v.

As stated earlier, ρ_m is the total density of the fluid: it is a function of C_v, W_w and the density for the tailings solids which is 2.82 t/m^3 in the considered TSF. With 50% of volume of water and 50% of solids we have for 1 m^3: 0.5 * 1 + 0.5 * 2.82 = 1.91 t/m^3 with 45% of solids we have 0.55 * 1 + 0.45 * 2.82 = 1.82 t/m^3.

The resulting rheological parameters for the Bingham module to simulate the "normal" ($C_v = 45$) and "high" ($C_v = 50$) tailings volumetric concentration scenarios are listed in Table 12.3.

In the example above, the selection of $C_v = 45$–50% to characterize "normal" and "high" tailings volumetric concentrations is key to the discussion.

The histograms in Figs. 12.2 and 12.3 must be considered with attention and the greatest care. Indeed, they present "gaps"; the medians and mean values seem too similar to correspond to some significantly different populations, etc.

In particular, the graph in Fig. 12.1, with mean at 45.14, standard deviation at 1.63 and max at 46.94 denotes either a sampling problem or some other ill-conditioned

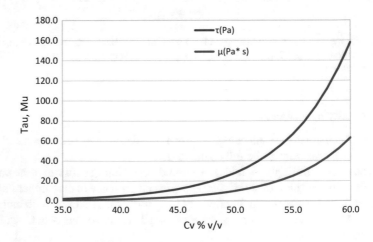

Fig. 12.3 Graph of τ_y and μ_m as a function of C_v

Table 12.2 τ_y and μ_m as a function of C_v

Cv (%v/v)	τ (Pa)	μ (Pa s)
35.0	2.1	0.6
36.0	2.5	0.8
37.0	3.0	0.9
38.0	3.5	1.1
39.0	4.2	1.3
40.0	5.0	1.6
41.0	5.9	1.9
42.0	7.1	2.3
43.0	8.4	2.8
44.0	10.0	3.3
45.0	*11.9*	*4.0*
46.0	14.1	4.8
47.0	16.7	5.8
48.0	19.9	6.9
49.0	23.7	8.3
50.0	**28.1**	**10.0**
51.0	33.4	12.0
52.0	39.7	14.5
53.0	47.2	17.4
54.0	56.1	20.9
55.0	66.7	25.1
56.0	79.2	30.2
57.0	94.2	36.3
58.0	111.9	43.7
59.0	133.0	52.5
60.0	158.1	63.1

Figures in italics indicate the "normal" (Cv =45); figures in bold indicate the "high" (Cv =50)

Table 12.3 Rheological parameters for the Bingham module to simulate the "normal" ($C_v = 45$) and "high" ($C_v = 50$) tailings volumetric concentration scenarios

Scenario	Cv (%v/v)	τy (Pa)	μm (Pa s)	ρm (t/m^3)
Normal tailings Cv	45	12	4	1.82
High tailings Cv	50	28	10	1.912

situation. It is indeed hard to believe that the maximum is $46.94 - 45.14 = 1.8$ distance to the mean when the standard deviation is 1.63, leading to 1.1 sigma bounds. That situation can't even be modelled with a J-shaped beta empirical distribution.

The graph shown in Fig. 12.2 has other blatant anomalies regarding its skewness.

Given the strong non-linearity of τ_y and μ_m, working with "arbitrarily selected average normal and high" C_v values from Figs. 12.1 and 12.2 is hazardous. Indeed this could lead to under- or over-estimates of flooding in the dam break analyses. Since there is a chance that a probabilistic analysis would lead to more favourable, reasonable parameters, it is worthwhile to devote some extra effort to it.

As both τ_y and μ_m are function of a single variable C_v (see Eqs. 12.1 and 12.2) we can get estimates for both variables using two point estimates of C_v using point estimates methods as shown in Table 12.4. Below are the P+, P− point estimates (see Rosenblueth 1975 for details) for the graph and values of Fig. 12.1:

$$P- = 45.14 - 1.63 = 43.51$$

$$P+ = 45.14 + 1.63 = 46.77$$

The interesting result brought by this example is that assuming "normal tailings" at the average value of $C_v = 45\%$ leads to underestimate the average value of τ_y and μ_m by at least 4%. Furthermore, given the large COV due to the non-linearity of the τ_y and μ_m functions, using the average value does not appear to be a reasonable selection.

Instead, considering the two variables and building four point estimates, would make it possible to perform a point estimate of the flood model and determine probabilistically the hazard zones.

One more important note. The rheology of a particular material can be highly type-dependent, and chemical and mineralogical properties can influence the tailings rheology. Thus a fluid's parameters can vary from one mine to another due to various reasons, including difference in ore and processing methods. Finally, the tailings

Table 12.4 Estimates for τ_y and μ_m using two point estimates of C_v using the point estimates methods (Rosenblueth 1975)

$C_v\%$			%		Evaluated standard deviation	Value for average	Variation %	C.O.V.%
Average	45.14	C_v+	46.77	0.4677				
Standard deviation	1.63	C_v-	43.51	0.4351				
$\tau+$	16.10	$\tau_{average}$	12.63			12.15	3.99	
$\tau-$	9.17	τ_{var}	12.00		3.46			27.43
$\mu+$	5.52	$\mu_{average}$	4.27			4.09	4.54	
$\mu-$	3.03	μ_{var}	1.55		1.25			29.15

rheology could change dramatically for changing tailings properties including degree of saturation, consolidation, and the increased presence of clay minerals.

12.2.2 Breach Size

Continuing with the dam/dykes example, the next step of a risk analysis involves the evaluation of breaches which lead to downstream consequences. The first step is of course to determine the breach size.

The literature shows that dam breach size models rely on many assumptions and are mostly based on geometry of the embankment and retained water levels/volumes (Franca and Almeida 2004; Morris 2005; Zagonjolli and Mynett 2005; USBR 1988).

In a 2013 study by Nourani and Mousavi (2013), 142 embankment dam breach data were collected from reliable references and dam breach equations analyzed. Dimensional analysis and multiple regression were used to predict maximum outflow from earth dam breach.

Uncertainty of empirical relations was determined using appropriate statistically method. The following general results were derived by (Nourani and Mousavi 2013) from collected data by studying 142 embankment dam breaches (Eq. 12.3):

$$2 * h_d \leq B_m \leq 3 * h_d \tag{12.3}$$

where B_m = average breach width (m); h_d = dam height (m); B_{top}/B_{bottom} = 1.13–1.64 width at top, bottom of the breach.

If we note as V_w the water volume above break point of bottom (m^3) and h_b the height of water above breach bottom the analyses performed in the study yielded the following regression (Eq. 12.4):

$$B_m = 2.2839 * V_w^{0.0635} * h_b^{0.8481} \quad \text{with } r = 0.918 \tag{12.4}$$

Please note that many of the necessary data are Space Observable or retrievable, by querying historic imagery databases.

The final step is to develop a breach outflow analysis and through that to evaluate the downstream consequences. Let's note that unless major changes intervene it is not necessary to redo a dam break analysis each time the risk assessment is updated, but variations in land use and density, which are all space observable will generate possible significant changes in consequences.

12.2.3 Outflow Volumes

The other input parameter is the outflow volume, for which many empirical relations exist, unfortunately each one of them proven wrong to some extent by actual failures, for various reasons.

Various researchers have attempted to define the volume of the outflow based on failures history and easy to determine dam characteristics. None of them is satisfactory, but they offer a "first stab" at evaluating runout volumes.

Rico et al. (2008) calculated the Volume of the runout VF using the total impounded volume (VT) in Mm^3 as in Eq. (12.5)

$$VF = 0.354 * VT^{1.01} \quad R^2 = 0.86 \tag{12.5}$$

and the outflow run-out distance travelled by the tailings in km (Dmax) is obtained using VF and the dam's height (in meters) at the time of failure (H) as in Eq. (12.6)

$$DMAX = 1.61 * (H * VF)^{0.66} \quad R^2 = 0.57 \tag{12.6}$$

Many analysts/engineers directly use such regression equations in a deterministic way to specify exposure. However, as site conditions vary significantly there is considerable uncertainty that needs to be quantified. Proof of that is the very low correlation calculated for Eq. (12.6). We will note that Eq. (12.5) basically states that 1/3 of the impoundment volume will spill (Tailings and water) through a dam breach.

Azam and Li (2010) analyzed 218 tailings accidents. Let's note that they divided that number of accidents by the number of "mines"—over 18,401—to publish an accident "frequency" of 1.2% for the hundred years of history they analyze, a quite absurd value considering that the number of mines does not equate the number of tailings or active tailings. Nevertheless they agree with our results (Oboni and Oboni 2013) that the frequency failures peaked in the 1970s–1980s and then declined afterwards, i.e., into the 2000s. Azam and Li stated that usually about one-fifth of the contained volume is released. However, they stated that the vast majority of releases were of unknown volume (see Fig. 6 of their referenced publication).

Case studies of past failures show the total outflow volume will often be less than the total storage above the bottom elevation of the breach because some portion of the tailings remains stable in the TSF under their own static strength. This is different than a water reservoir, where most or all of the stored water above the bottom of the breach elevation will out flow.

None of the above accounts for liquefaction (see Sect. 9.3) and Rico et al. (2008) rightly pointed out that some of the parameters contributing to the uncertainty in the predictions include sediment load, fluid behaviour (depending on the type of failure), topography, the presence of obstacles stopping the flow, and the proportion of water stored in the tailings dam (linked to meteorological events or not).

A paper entitled "Predicting Tailings Dams Breach Release Volumes for Flood Hazard Delineation" (O'Brien et al. 2016) attempts to give solutions to the same problem).

It uses again a set of past failures (with all the problems we have discussed earlier) and derives three relationships (Eqs. 12.7–12.9) based on the breached dam height as follows:

$$V_{upper\ bound} = 20,419 * H^{1.2821} \tag{12.7}$$

$$V_{mean} = 3604 * H^{1.2821} \tag{12.8}$$

$$V_{lower\ bound} = 1052 * H^{1.2821} \tag{12.9}$$

As most of the data used by O'Brien et al. were the same than those used by Azam and Li (2010) and most of the past cases were low-volume, low-height dams, we would advise to use, for dams above 45–50 m and out of prudence, the upper bound solution.

Lately a research paper entitled "Tailings Dams Failures: Updated Statistical Model for Discharge Volume and Runout" by Concha Larrauri and Lall (2018) has lead these authors to more detailed discussions and to publish a free-access app[1] which has the capability of computing the probability of exceeding a runout volume specified by the user based on a set of historic failures relationships. In this manner, the uncertainty regarding the volume can be considered when estimating the runout distance. The app also provides various percentiles of the results.

Concha Larrauri and Lall used 29 failures instead of the original 22 used by Rico et al. At the end of the day they also show that the total runout volume is appx. 1/3 of the impoundment volume. They state that this type of analysis should always be considered probabilistically due to the uncertainties. We believe that introducing in the sample population failures like Samarco, where tailings were carried hundreds of kilometres away by the Rio Doce (river transport mechanism), constitutes a serious flaw intrinsic to the dataset. In fact, although tailings travelled more than 600 km (due to the river), 90% of the released volume remained within 120 km from the source (which is already quite significant as well).

For Samarco the median prediction of runout volume by Concha Larrauri and Lall was approx. 50% of the actual release whereas for Mount Polley the median prediction of runout volume was approx. 85% of the actual release. Additionally, for smaller TSF the margin of error can become very large.

Rourke and Luppnow (2015) have recently published "The Risks of Excess Water on Tailings Facilities and Its Application to Dam-break Studies", which correlated recent catastrophic failures which reached the bottom of the TSF to the percentage of the TSF surface inundated. The resulting correlation is staggering, as shown in Fig. 12.4.

---

[1] https://columbiawater.shinyapps.io/ShinyappRicoRedo/.

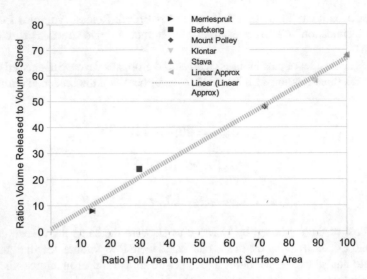

Fig. 12.4 Predicted tailings outflows volume ration versus impoundment area ratio (Rourke and Luppnow 2015)

Although the correlation seems enticing, a word of caution must be formulated. In fact, this correlation does not include any information on the surface, volume and depth of the TSF. For example, the volume of Mount Polley was significantly larger than all the other cases. Thus, it could be that this wonderful correlation is a "fluke" and many more cases will have to be brought into the analysis to develop confidence.

12.2.4 Flood Hazard

Once the breach size, the rheology and the outflow volume have been estimated a model can be built to evaluate the flooding. There are several well-known commercial software programs that can help to carry out the numerical analyses. At the end, the flood is described point by point in terms of water depth and velocity.

Smith et al. (2014) have formulated a depth (m)–velocity (m/s) chart making it possible to define the flooding hazard at each point reached by the flood. This is shown, for reference, in Fig. 12.5.

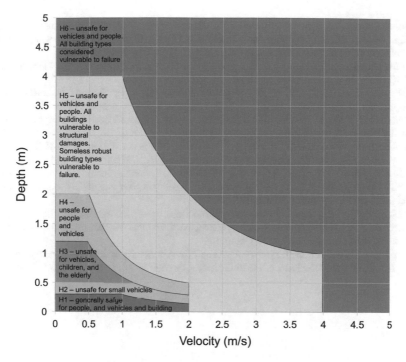

Fig. 12.5 Smith et al. (2014) depth (m)–velocity (m/s) chart to evaluate flooding hazard

References

[AGSLT 2007] Australian Geomechanics Society Landslide Taskforce (2007) Practice Note Guide-lines For Landslide Risk Management, Landslide Practice Note Working Group, Journal and News of the Australian Geomechanics Society

Azam S, Li Q (2010) Tailings Dam Failures: A Review of the Last One Hundred Years, Waste Geo Technics

Darbre G (1998) Dam risk analysis. Report, Federal Office for Water and Geology. Dam Safety, Bienne

Franca MJ, Almeida AB (2004) A computational model of rockfill dam breaching caused by over-topping (RoDaB), Journal of Hydraulic Research 42(2): 197–206

Julien PY (2010). Erosion and Sedimentation, 2nd ed., Cambridge, Cambridge University Press

Larrauri Concha P, Lall U (2018) Tailings Dams Failures: Updated Statistical Model for Discharge Volume and Runout *Environments* 5(2), 28; https://doi.org/10.3390/environments5020028

Morris, M. W. (2005) IMPACT: Final technical report, Proc. of the 1st IMPACT Workshop, HR Wallingford, UK

Nourani, V, Mousavi, S (2013) Evaluation of Earthen Dam-Breach Parameters and Resulting Flood Routing Case Study: Aidoghmosh Dam, International Journal of Agriculture Innovations and Research, Volume 1, Issue 4, ISSN (Online) 2319–1473

Oboni C, Oboni F (2013) Factual and Foreseeable Reliability of Tailings Dams and Nuclear Reactors -a Societal Acceptability Perspective, Tailings and Mine Waste 2013, Banff, AB, November 6 to 9, 2013

O'Brien J, Gonzalez-Ramirez N, Tocher R, Chao K, Overton D (2016) Predicting Tailings Dams Breach Release Volumes for Flood Hazard Delineation. *Annual conference proceedings*/Association of State Dam Safety Officials, vol. 1, pp. 207–224

Rico M, Benito G, Diez-Herrero A (2008) Floods from tailings dam failures. J. Hazard. Mater. 154(1–3):79–87

Rourke H, Luppnow D (2015) The Risks of Excess Water on Tailings Facilities and Its Application to Dam-break Studies. Tailings and Mine Waste Management for the 21st Century, Sydney, NSW https://www.in.srk.com/files/au/The_Risks_of_Excess_Water_on_Tailings_Facilities_and_Its_Application_to_Dam-break_Studies.pdf

Rosenblueth E (1975) Point Estimates for Probability Moments. Proceedings of the National Academy of Sciences of the United States of America PNAS 72(10): 3812–3814

Smith GP, Davey EK, Cox RJ (2014) Flood Hazard, WRL Technical Report 2014/07, UNSW Water Research Laboratory https://knowledge.aidr.org.au/media/2334/wrl-flood-hazard-techinical-report-september-2014.pdf

[USBR 1988] U.S. Bureau of Reclamation (1988) Downstream hazard classification guidelines, ACER Technical Memorandum No. 11, Assistant Commissioner-Engineering and Research, U.S. Department of the Interior, Denver, Colorado

Zagonjolli M, Mynett AE (2005) Dam breach analysis: A comparison between physical, empirical and data mining models, Proc. of the 29th IAHR Congress Seoul, South Korea, 753–754

Chapter 13
Tolerance and Acceptability

In this chapter we discuss risk tolerance and acceptability, how they used to be defined, how they can be defined. Tolerance/acceptability thresholds have to be developed independently from risks to ensure unbiased results.

What constitutes acceptable/tolerable risk depends on the perspectives (https://www.riskope.com/2013/11/07/social-acceptability-criteria-winning-back-public-trust-require-drastic-overhaul-of-risk-assessments-common-practice/) of the organizations or persons involved. A family or community living in the dam breach inundation zone will have different perspectives than the mining companies whose executives have been told that their designs and risk management practices have been based on best practices. The level of risk deemed acceptable to a corporation is not necessarily what may be considered acceptable to government or the public (https://www.riskope.com/2011/09/06/what-fukushima-2010-nuclear-accident-the-twin-towers-911-terror-attack-deadly-traffic-accidents-and-aquila-earthquake-italy-hurricanes-have-in-common/), even if it is based on a collaborative engagement process. The tolerance and acceptability thresholds are set by society: government, authorities, regulators, or others who are directly concerned (Darbre 1998). The acceptability of risk also has to consider closure implications as well the broad benefits to society that flow from economic development. The level of protection (mitigation) to be applied to any system is therefore a function of the comparison between the various risk scenarios and the risk acceptability/tolerance threshold.

Two thresholds should be derived: an acceptability threshold and a tolerance threshold. The acceptability threshold denotes a rejectable quality and is always lower than the tolerance threshold, which is an upper boundary of permissible deviation (Kalinina et al. 2016). These threshold can be set either in the form of constant values or curves.

© Springer Nature Switzerland AG 2020
F. Oboni and C. Oboni, *Tailings Dam Management for the Twenty-First Century*,
https://doi.org/10.1007/978-3-030-19447-5_13

Here modern risk assessment methods help immensely to see through the apparent complicated series of scenarios and tolerances and then make a balanced decision based on a meaningful public engagement process. From a corporate perspective, the financial consequences of a tailings dam failure or the possible malfunctioning of thickened, filtered or paste alternatives all carry significant financial consequences. The problem is that the possible alternatives to slurry deposition have not yet created the same body of knowledge that could support development of professional guidances and protocols of a quality equal to that for slurry deposition. Thus it is paramount that the risk assessment used to compare the alternatives can explicitly tackle the different levels of uncertainties and lead to a result that should be approximately right, rather than precisely wrong.

If a project or an alternative has too many intolerable risks (i.e., the aggregated intolerable risk is too high), it may be time to drop the project/alternative rather than build it nevertheless and try to mitigate.

From a public perspective, what defines acceptable risk is not found in the results of a dam safety review report. As an example, the guideline Legislated Dam Safety Reviews (APEGBC 2014) states:

> The determination of what is the acceptable level of risk or safety for the various elements which are identified as being at risk is not the role of the qualified professional engineer and is outside the scope of the dam safety analysis. The acceptable level of risk must be established and adopted by the regulatory authority in consultation with the dam owner. However, an assessment of the various elements at risk, through the dam failure consequences classification established by the relevant regulatory authority will guide the qualified professional engineer's dam safety analysis.

What this basically states is that a government's or regulatory authority's approval of a design, based on an acceptance of its identified risks, defines what constitutes an acceptable level of risk. The assurance statement required as part of the Dam Safety Review Report verifying that "the dam is reasonably safe" means nothing more than the dam's level of risk is no worse than that level of risk previously approved by the government. Clearly the government is the final arbiter as to the determination of acceptable risk on a case by case situation. The question is: what is the responsibility of the government in this case?

Voluntary risk can be defined as the risk proceeding from the will or from one's own choice or consent. Involuntary risks are generated by situations contrary to or without choice. In our litigation-prone societies, the distinction becomes fuzzier every day and is linked to the knowledge of the existence of the hazard. A perfectly ignorant person can only have involuntary risks, as this person cannot formulate any choice based on potential hazard exposure. At the other end of the spectrum a perfectly knowledgeable person only has voluntary risks. Having voluntary risks and not mitigating them can be considered criminal negligence, thus leading to the paradox of good companies and individuals hiding the fact they perform good risk assessments, in order to avoid exposure to legal proceedings.

When discussing human acceptability a distinction will be made between location-based risks and societal risks. Location-based risks derive from the annualized likelihood of a person being killed at a given location as a direct result of an accident associated with hazardous activities undertaken there. Societal risks represent the likelihood of a group of people who are not directly engaged in an activity involving a hazardous substance being killed in an accident arising out of that activity. Location-based risk is an expression of the risk that someone who lives or works in a place where a hazardous activity takes place is exposed to. Societal risk is quite different from location-based risk: it looks at the consequences of mishaps from the very broad point of view of an entire society, possibly physically and emotionally removed from the mishap itself; as such, it is of interest mainly to public administrators (Geerts et al. 2016).

> The determination of the acceptable risk level generally differs depending on whether a phenomenon is natural or man-made, or a private or a public (societal) issue.

Acceptable risk and corporate commitment (https://www.riskope.com/2010/06/08/bp-crisis-rational-analysis-what-bp-did-not-perform/) are inextricably linked. A company that is able to demonstrate its commitment to the public through the application of a strong tailings governance framework will stand a better chance of having its proposed tailings plans accepted by the public and maintain its SLO. A company that is able to demonstrate its commitment to its own employees will more likely develop a lower risk proposal. A company that is able to demonstrate its commitment to the regulatory authorities and the public will find the permitting process much easier to navigate. A company that has done all these things will also stand an excellent chance of not experiencing a catastrophic dam failure.

The ideal outcome is to have all parties agree, based on informed opinions, that the risks, and their mitigating measures, for a proposed mine plan are acceptable. Informed opinions are only possible when all parties, particularly the public, have been provided with:

- the opportunity to participate in a meaningful communication and engagement process which clearly states what mitigations will be implemented, by whom and when, with a clear roadmap and accountability;
- the consequence rating of the proposed dam, including the results of a dam breach and inundation study and an unbiased risk assessment;
- information that supports the selection of the deposition method and site location;
- information that demonstrates beyond reasonable doubt that the owner is committed to the management of the dam and its critical risks through the establishment of a comprehensive management framework and assurance program; and
- information that demonstrates that the government has established and will be committed to an effective compliance and enforcement regime.

13.1   Historic Tolerance Thresholds

13.1.1   Examples of Constant-Value Acceptable and/or Tolerance Thresholds

Looking back in time various authors (Wilson 1984; Comar 1987; Renshaw 1990) defined simple societal constant-value risk acceptability criteria, as shown in Table 13.1.

For landslides the Australian Geomechanics Society Suggested tolerable loss of individual life risk as shown in Table 13.2 (AGSLT 2007).

13.1.2   Examples of Acceptable- and/or Tolerance-Threshold Curves

Countless curves have been developed for different industries/hazards. In log-log scale these curves appear as straight lines, thus "in real life" they have a hyperbolic shape. Let's immediately remark that in a log-log graph a straight line of slope -1 ($-45°$, i.e., going from $p = 10^{-4}$ to C $= 10,000 = 10^{+4}$) corresponds to a centred hyperbolic function of constant risk. That would be the tolerance curve of a "rational" being who does not feel any emotion toward small but extremely likely losses (the "pain in the neck" type of maintenance nuisance) or very large losses at very low likelihood (say the Fukushima type of phenomenon). As the mathematics have not

Table 13.1 Historic societal constant value risk acceptability criteria

Author	Units	Unacceptable risk/upper bound	Negligible risk/lower bound
Renshaw	Probabilities of fatality of one individual per year of exposure to the risk	10^{-5}	10^{-7}
Comar		10^{-4}	10^{-5}
Wilson		10^{-3}	10^{-6}
Renshaw	Probabilities of fatality of ten individuals per year	10^{-5}	10^{-7}

Table 13.2 AGS suggested tolerable loss of life individual risk

Situation	Suggested tolerable loss of life risk for the person most at risk per year
Existing slope/existing development	10^{-4}
New constructed slope/new development/existing landslide	10^{-5}

changed over time some passages below are strongly inspired from *Improving Sustainability through Reasonable Risk and Crisis Management* (https://www.riskope.com/knowledge-centre/tool-box/books/) (Oboni and Oboni 2007).

The straight line in the log-log plot was called the boundary Line in early literature (Farmer 1967; EDI 1989). As per the exponent b, a value smaller than -1 indicates risk averseness whereas b larger than -1 indicates risk appetite (Plattner 2005). Generally human beings have an appetite for voluntary risks, but an averseness to involuntary risks. Moreover, many Perception Affecting Factors (PAF)[1] enter into the delicate equilibrium between risk averseness and appetite, as described in Table 13.3.

Table 13.3 The delicate equilibrium between risk averseness and appetite

PAF	Represents
Voluntariness	Voluntariness
Reducibility	Reducibility, predictability, avoidability
Knowledge	Familiarity, knowledge about risk, manageability
Endangerment	Controllability, number of people affected, fatality of consequences, distribution of victims (spatio-temporal), scope of area affected, immediacy of effects, directness of impact
Subjective measure of extent of damage	Extent of damage
Subjective measure of frequency of event	Frequency of event

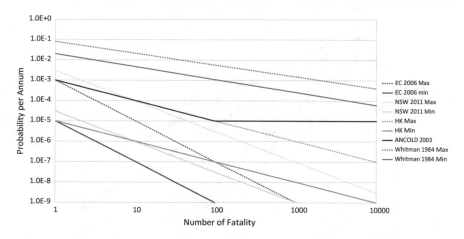

Fig. 13.1 Comparison between various risk tolerance thresholds, consequences expressed in fatalities. Notice that the vertical axis (probabilities per annum) extends well below the credibility threshold for a number of examples

---

[1] Whereas Plattner (2005) considers that perceived risks to be generally higher than real, objective risks, we take the stance of possible overestimation and possible underestimation of risks, as factors as familiarity, knowledge, controllability etc. blatantly lead people to underestimate risks they are exposed to.

Well-known examples of log-log straight thresholds are found in Christou et al. (2006), NSW (2011), Hong Kong Government (1994), ANCOLD (2003) and Whitman (1984). They are depicted in Fig. 13.1.

By simply reasoning on the slope of the graphs of Fig. 13.1 one can immediately see that:

- Whitman's (1984) tolerance thresholds slopes depict a risk-prone society;
- Hong Kong (1994) curves are ISO-risk;
- NSW (2011) curves are risk-averse;
- EC (Christou et al. 2006) curves are strongly risk-averse,
- ANCOLD first depict an ISO-risk behaviour and then, from 100 casualties onward, depict a risk-indifferent behaviour.

The Whitman tolerance is depicted in Fig. 13.2 together with data from accidents of the times it was defined. In the graph in the original paper (Whitman 1984), reproduced here as is, the "cost of a life" was equated to one million dollars with no further emotion or discussion. That was considered just fine in those times. Today we do not attribute a cost to human life, a notion often considered repugnant and ethically unacceptable, but rather, we look at accepted capital expenditure in various countries to save the life of a citizen potentially exposed to natural hazards to find a "bounding value". In other words, this book uses the mitigative investment a society

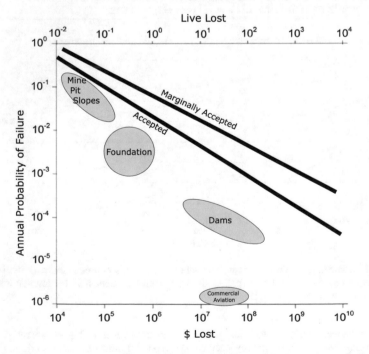

Fig. 13.2 Whitman tolerance data from accidents of those times. NB: "cost of casualty" assumed to be 1 M USD

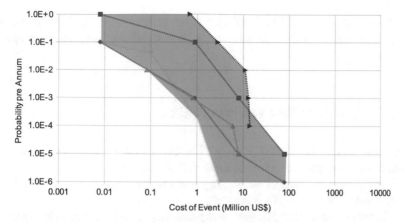

Fig. 13.3 Empirically (facilitated workshops) developed risk-tolerance thresholds for a set of medium to large corporations. The envelope is displayed in blue shade

is ready to make to save a life or its "willingness to pay" (WTP) attitude (Marin 1992; Mooney 1977; Jones-Lee 1989; Lee and Jones 2004; Pearce et al. 1996).

13.1.3   Examples of Monetary Acceptable and/or Tolerance Thresholds Curves

For a monetary-only tolerance threshold the curves have a totally different appearance. The reason is that in the event of a certain consequence being realized, the company will fail, or lose its freedom, no matter what. It is therefore modelled as an activation function. For a set of clients we developed the set of tolerance thresholds shown in Fig. 13.3. The curves are bundled to preserve confidentiality.

13.2   Modern Risk Tolerance

Risks can be sorted by decreasing value and a prioritized list can be delivered. However that prioritization is far from "best practices" as it lacks a vital piece of information: what risks actually do have the potential to "hurt"? It is also vital for the client to distinguish between high-probability low-consequence events and their opposite, i.e., low-probability, high-consequence events, something that common PIGs do not allow, simply because their colouring schemes have been arbitrarily defined, without proper understanding of their meanings. Indeed, in many common practice approaches (FMEA, risk matrices) one can see that a Fukushima-type of event receives the same colour as, say, the next migraine of the plant director, which is obviously a source of ill-conditioned behaviours and conclusions. In order to perform sensible prioritization, risks must be compared to corporate and societal risk tolerance.

13.2.1   Corporate Risk Tolerance (CRT)

Corporate Risk Tolerance (CRT) is a "business" function linking the financial consequences a corporation views as a tolerable losses at various levels of probability. Thus CRT should be seen primarily as a perceived tolerance risk threshold, the real financial limit being generally higher.

CRTs are unique to each corporation, operation, project and timing (within the life cycle) and can be determined through facilitated questions/answers using a methodology we have developed over the years, after developing "point-by-point" curves in workshops with boards of directors around the world and looking at examples developed by others.

In recent decades "standardized levels of risk reduction" have been formulated, and at least three of these definitions are now in common use among analysts. These three levels of risk mitigation also represent a convenient way to elude explicit tackling of risk acceptability, especially when the delicate theme of human life has to be dealt with.

ALARA is an acronym for the phrase "as low as reasonably achievable" (Wilson and Crouch 1982). It is most often used in reference to chemical or radiation exposure levels.

ALARP stands for "as low as reasonably practicable", and is a term often used in the milieu of safety-critical and high-integrity systems. The ALARP principle is that the residual risk shall be as low as reasonably practicable (Wilson and Crouch 1982). For a risk to be ALARP it must be possible to demonstrate that the cost involved in reducing the risk further would be grossly disproportionate to the benefit gained. The ALARP principle arises from the fact that it would be possible to spend infinite time, effort and money attempting to reduce a risk to zero. It should not be understood as simply a quantitative measure of benefit against detriment. It is more a best common practice of judgment of the balance of risk and societal benefit.

BACT (best available control technology). For example: an emission limitation based on the maximum degree of emission reduction (considering energy, environmental, and economic impacts) achievable through application of production processes and available methods, systems, and techniques. BACT does not permit emissions in excess of those allowed under any applicable Clean Air Act provisions. Use of the BACT concept is allowable on a case-by-case basis for major new or modified emissions sources in attainment areas and applies to each regulated pollutant.

13.2.2 Societal Risk Tolerance

Societal risk tolerance is generally based on a "casualty cost" of failures. Figure 13.4 shows the ANCOLD tolerance criteria as well as the Whitman life risk tolerance compared with what occurred in the decades 1979 and 1999to TSF and Nuclear 5+ accidents over 14,000 years-reactors (Oboni and Oboni 2013). Recent studies have shown that Whitman thresholds remain valid nowadays provided a correct WTP is used (see Sect. 13.1.2).

Once the risks incurred by an operation are estimated, rational and sustainable decisions on risk mitigation are generally requested by clients wishing to adopt risk management methods and maximize the investment they have made by performing a risk assessment. These can only be taken after an explicit risk tolerance function in defined. The tolerance function can be:

- derived formally (mathematically) from client's financial data;
- defined empirically;
- derived from public opinion tests;
- "negotiated".

Generally the final tolerance curves selected by clients are the result of a mix of these various approaches.

Tolerable risk curves are always project- and owner-specific and indicate the level of risk which has been deemed acceptable by an owner for a specific project or operation (possibly taking into account public opinion). This means, as an example, that within large companies corporate risk tolerability may differ quite substantially from that of a branch operation.

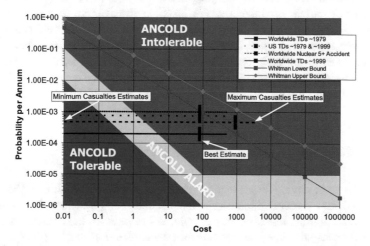

Fig. 13.4 ANCOLD tolerance criteria, Whitman life risk tolerance compared with TSF failure occurrences in the decades around 1979 and 1999, Nuclear 5+ accidents over 14,000 years-reactors (Oboni and Oboni 2013)

Risks which plot to the left and below the tolerance curve are deemed bearable. Risks which plot to the right and above the curve are deemed unbearable and some measures of mitigation are considered necessary to reduce their likelihood. Reducing the likelihood of an impact may be, for example, as simple as imposing "no stop" zones on a road.

When working empirically, two curves should be developed, one representing the optimistic, the other the pessimistic view of tolerance. The area between the curves represents a range of uncertainty on tolerance defined by an organization. When data are available theoretical curves can be developed and then discussed with key personnel.

> **Why is risk tolerance so important? One example related to a classic trend: over-estimating outcome severity after one mishap.** Let's consider a system that causes, on the average, one accident every one hundred years. Most of these accidents have relatively small consequences, say one fatality for each. Once in a while there may be a catastrophic event generating ten fatalities. If the catastrophic event happens to occur, the public (or regulatory agencies) may believe that all accidents have catas-trophic outcomes, thus they demand more safety measures than are justified by the actual damage expectation. Such a claim is not restricted to the particular facility that caused the accident; improvements are required for all other facilities of this type.

The development of empirical-estimated tolerance curves requires caution and continuous calibration as the extent of correlation between an individual's estimate or ranking of probabilities and the true value/ranking is usually quite weak, sometimes even in the order of zero (Gordon 1924; Peterson and Beach 1976). However, it has been demonstrated that pooled judgments correspond better with the truth as the number of individuals increases. For instance, the average correlation between individual judgments and the correct rank order may increase twofold when pooled across seven individuals, and twenty individuals may have an excellent ranking. No wonder juries are made out of twelve people! Jokes aside, this is one of the reasons why risk assessments, and in particular risk tolerance curves, should always be defined by a group, and not by an individual (Hofstätter 1986; Wilde 2001).

13.2.3   Tolerance Versus Time and Tolerance Zones

Risk tolerance is obviously a function of an organization's wealth. In the case of some industries—extractive industries in particular—this translates into a function of time. In the mining industry, for example, as ore reserves are depleted, tolerance decreases because the company has less future wealth to buffer a hit (the operational safety margin decreases with time) (Fig. 13.5).

Corporate wealth also has to do with the attitude a corporation may have in defining risk tolerance. Figure 13.6 shows in orange a corporate-developed tolerance

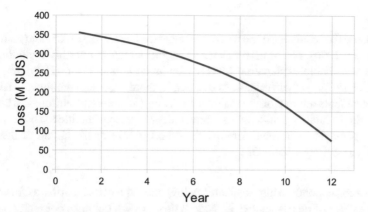

Fig. 13.5 Decrease of risk tolerance at a mine as reserves deplete over time

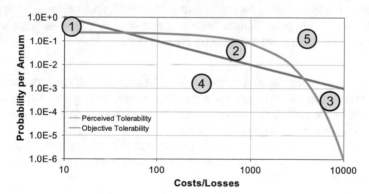

Fig. 13.6 The orange curve is a corporate developed tolerance threshold compared to a blue "emotionless" objective tolerance threshold

threshold compared to a "emotionless" objective tolerance threshold (blue line in log-log graph). As can be seen zones 1 and 3 are below the objective line but above corporate tolerance, while zone 2 is corporately tolerable, but above the objective line. Zones 4, 5 are respectively entirely tolerable, entirely intolerable, thus they do not necessitate detailed discussion.

In order to explain why the differences of perception (zones 1, 2, and 3) occur, let's look at an example based on parking violations fines:

- If there is a very high likelihood of being caught in illegal parking, even if the fine is modest, most people will be cautious (Zone 1).
- If the fine is substantial, and there are very few police patrolling parking, most people will also be cautious (Zone 3).
- In-between, most people will take a chance! (Zone 2; in this case not sure we are talking about human evolution!).

And now let's describe the three zones.

Zone 3

This zone is the result of a trend to overestimate frequency and outcomes (Kumamoto and Henley 2000), resulting in risk aversion for outcomes possibly including fatalities. A reasonable explanation of this phenomenon was given (Bohnenblust and Schneider 1987) in Switzerland. According to them, poor estimations of risks after severe accidents are one of the major reasons for risk-aversive attitudes, which prefer ten single-fatality accidents to one accident with ten fatalities. Thus, risks can be mis-estimated or overestimated with respect to size or likelihood of outcomes, generally resulting in Zone 3.

Zone 1

This area corresponds to high-probability low-consequence scenarios, as for example having a piece of equipment which breaks down often, but does not really have any significant consequence: despite the fact that the risks are lower than the constant risk curve, clients will often decide to change the equipment, because the impact is overestimated and/or by a "fatigue" process: this is often voiced as "that piece of equipment was a pain in the neck, so we changed it!".

Zone 2

This is the area where we find ourselves in a "false comfort zone" where we tend to play with likelihood and consequences, often overestimating our tolerance: when consequences are neither too heavy nor too frequent, we do love to take risks, and this indeed is one of the engines of evolution.

To conclude, it is precisely in defining the tolerance curve in Zone 2 that the most care has to be exerted by the risk assessments team, especially when considering the influence of PAFs (Table 13.3).

References

[AGSLT 2007] Australian Geomechanics Society Landslide Taskforce (2007) Practice Note Guidelines For Landslide Risk Management, Landslide Practice Note Working Group, Journal and News of the Australian Geomechanics Society

[APEGBC 2014] APEGBC 2014. Association of Professional Engineers and Geoscientists of British Columbia (2014) Professional Practice Guidelines – Legislated. Dam Safety Reviews in BC V2.0. https://www.apeg.bc.ca/For-Members/Professional-Practice/Professional-Practice-Guidelines

[ANCOLD 2003] Australian National Committee on Large Dams (2003) Guidelines on risk assessment. Sydney: ANCOLD

Bohnenblust H, Schneider T (1987) Risk appraisal: Can it be improved by formal decision models? In: Uncertainty in Risk Assessment, Risk Management, and Decision Making, ed. VT Covello, et al., pp. 71–87, New York, Plenum Press

Comar C (1987) Risk: A pragmatic de minimis approach, in: De Minimis Risk, ed C. Whipple, pp. xiii–xiv, New York, Plenum Press

Christou M, Struckl, M, Biermann, T (2006) Land use planning guidelines in the context of article 12 of the Seveso II Directive 96/82/EC as amended by Directive105/2003/EC. European Commission Joint Research Centre, Institute for the Protection and Security of the Citizen. EUR 22634 EN FR DE

Darbre G (1998) Dam risk analysis. Report, Federal Office for Water and Geology. Dam Safety, Bienne

[EDI 1989] Eidgenössisches Departement des Innern (1989) Verordnung über den Schutz vor Störfallen (Störfallverordnung, SFV), Entwurf, Bern

Farmer F (1967) Siting criteria – a new approach, in: Containment and siting of nuclear power plants, pp. 303–329, Vienna, International Atomic Energy Agency (IAEA)

Geerts R, Heitinka J, Gooijerb L, van Vlietb A, Scheresa R, de Boerc D (2016) Societal Risk and Urban Land Use Planning: Creating Useful Pro-Active Risk Information. CHEMICAL ENGINEERING TRANSACTIONS VOL. 48, The Italian Association of Chemical Engineering Online at www.aidic.it/cet

Gordon K (1924) Group judgments in the field of lifted weights. Journal of Experimental Psychology 7: 398–400

Hofstätter P (1986) Gruppendynamik, Hamburg, Rowohlt

Hong Kong Environmental Protection Department (1994) Practice note for professional persons. ProPECC PN 2/94 https://www.epd.gov.hk/epd/sites/default/files/epd/english/resources_pub/publications/files/pn94_2.pdf

Jones-Lee MW (1989) The Economics of Safety and Physical Risk, Oxford, Blackwell

Kalinina A, Spada M, Marelli S, Burgherr P, Sudret B (2016) Uncertainties in the risk assessment of hydropower dams state-of-the-art and outlook, Zurich, ETHZ https://www.ethz.ch/content/dam/ethz/special-interest/baug/ibk/risk-safety-and-uncertainty-dam/publications/reports/RSUQ-2016-008.pdf

Kumamoto H, Henley EJ (2000) Probabilistic Risk Assessment and Management for Engineers and Scientists, 2nd ed. New York, Wiley-IEEE Press

Lee, E.M., Jones, D.K.C., Landslide Risk Assessment, Thomas Telford, 2004

Marin A (1992) Costs and Benefits of Risk Reduction. Appendix in Risk: Analysis, Perception and Management, Report of a Royal Society Study Group, London

Mooney GM (1977) The Valuation of Human Life, Macmillan, New York

[NSW 2011] NSW Government: Hazardous Industry Planning Advisory Paper No 4 Risk Criteria for Land Use Safety Planning, January 2011. https://www.planning.nsw.gov.au/-/media/Files/DPE/Other/hazardous-industry-planning-advisory-paper-no-4-risk-criteria-for-land-use-safety-planning-2011-01.pdf?la=en

Oboni F, Oboni C (2007) Improving Sustainability through Reasonable Risk and Crisis Management, ISBN 978-0-9784462-0-8

Oboni C, Oboni F (2013) Factual and Foreseeable Reliability of Tailings Dams and Nuclear Reactors -a Societal Acceptability Perspective, Tailings and Mine Waste 2013, Banff, AB, November 6 to 9, 2013

Pearce, DW, Cline WR, Achanta AN, Fankhauser S, Pachauri RK, Tol RSJ, Vellinga P (1996) The Social Costs of Climate Change: Greenhouse damage and the benefits of control, in: Climate Change 1995: Economic and Social Dimensions of Climate Change. Contribution of Working Group III to the Second Assessment Report of the IPCC, Cambridge, Cambridge University Press

Peterson CR, Beach LR (1976) Man as an intuitive statistician. Psychological Bulletin, 68: 29–46

Plattner T (2005) Modeling public risk evaluation of natural hazards: a conceptual Approach, Natural Hazards and Earth System Sciences 5: 357–366

Renshaw FM (1990) A Major Accident Prevention Program, Plant/Operations Progress 9(3): 194–197

Whitman RV (1984) Evaluating calculated risk in geotechnical engineering. J. Geot. Engineering 110(2): 145–188

Wilde GJS. (2001) Target risk 2, Toronto, PDE Publications

Wilson R (1984) Commentary: Risks and their acceptability, Science, Technology, and Hitman Values 9(2): 11–22

Wilson AC, Crouch E (1982) Risk/Benefit Analysis, Cambridge MA, Ballinger Publishing Company

Chapter 14
Risk Assessment for the Twenty-First Century

This chapter delves with the methodological details of modern risk assessments in compliance with the "specs" defined in Part I, and in particular Chap. 6.

In response to government and public expectations (see Part I) there is an increasing requirement for the assessment of alternate deposition methods for the purpose of reducing site-specific risks and impacts. As stated in the Australian Government Tailings Management publication (AG TMH 2016):

Regulators nowadays expect all TSF design submissions to demonstrate beyond reasonable doubt that sustainable outcomes will be achieved during operations and after closure by the application of leading practice risk-based design that:

- Fully assesses the risks associated with tailings storage at a particular site;
- Compares the suitability of all available tailings storage methods, in particular those that involve tailings dewatering and/or eliminate the requirement for the damming of surplus water within the TSF;
- Demonstrates that the tailings storage method selected will manage all risks to within acceptable levels and as low as reasonably practicable (ICOLD 2013).

We might add that leading practice risk-based design also:

- considers closure requirements and its associated risks;
- enables the public to participate in a collaborative manner in the examination of alternate deposition methods and their related risk management strategies.

Because of their sizes, their societal importance and the potentially catastrophic consequences of their failure, it is essential to assess the risks generated by large dams rigorously (Brown 2017). Their management should be proactive at all stages of

© Springer Nature Switzerland AG 2020
F. Oboni and C. Oboni, *Tailings Dam Management for the Twenty-First Century*,
https://doi.org/10.1007/978-3-030-19447-5_14

their lives so that the societal risk of dam failure remains within the ALARP zone (ANCOLD 2003; Barker 2011) (see Sect. 13.2.1).

Regarding permit conditions, leading practice requires that permits and permit amendments be granted on the basis of strict conditions related to critical operating parameters and risk mitigating strategies and that they be measurable and enforceable.

As we have stated earlier, there is a blatant gap between the above requirements, similar to those discussed in Chap. 5, and the capabilities of common practices (see Chap. 6 and Sect. 7.8). Pigs do not fly (https://www.riskope.com/wp-content/uploads/2013/06/Riskope-White-Paper.pdf) when tailings systems need to be evaluated at any time horizon (Oboni et al. 2013). Furthermore Canadian requirements such as NI43-101 (https://www.riskope.com/2017/05/10/technical-session-towards-improving-environmental-social-disclosure-ni43-101/) will force mining industries to develop better risk assessments which allow potential investors to UNDERSTAND realistic risk landscape of mining corporations.

Thus, in this present section we examine how to build an ISO 31000-compliant modern quantitative risk approach which is scalable, drillable, convergent and allows Bayesian updates as new data become available. In other words, a risk assessment which covers the specs indicated in Chaps. 5 and 6. Furthermore, risk assessment should be used for RIDM as we will show in Part III.

We also discuss the use of risk tolerance (thresholds discussed in Chap. 13), which should be developed independent of the probability consequence register in order to avoid biases and "conflicts of interest". Incidentally we will show that families of risk (i.e., tolerable risks, intolerable, but manageable and intolerable and unmanageable) will become self-evident once tolerance is used. We will show that this latter family covers the "strategic risks" and that these are the result of a risk assessment and not an arbitrary, a priori, intuition based on "the scariest stuff".

Finally we will explain the principle of the quantitative, scalable, convergent, drillable risk assessment methodology, which transparently includes uncertainties and which we deploy through the array of mining and non-mining cases we study in our day-to-day practice.

Let's define the terms we have just used and are paramount to define a modern, sustainable, rational and prone to support risk informed decision-making risk assessment methodology:

- Uncertainties exist in both the probability and the consequences of events (Sect. 7.4). They are included using ranges based on the state of knowledge at the moment of the deployment. These will be altered during the life of the system (see "Updatable" below).
- Convergent means that all hazards potentially present (technical, man-made, natural, etc.) (Chaps. 9 and 12) are considered in the aggregate risk evaluations.
- Scalable means that the risk register is built in such a way that new projects, alterations, in-depth studies of some structures or system's elements can be added and developed within the same risk register.

- Updatable means that at any time the values of probabilities and consequences (Sect. 10.1.2) as well as their respective uncertainties can be made current on the basis of data as it is obtained, delivering a new risk landscape of the company.
- Drillable means that data can be retrieved using various queries, such as asking which are the highest (tolerable, intolerable) risks within the company generated by potential earthquake, or employee dishonesty.

14.1   ISO 31000

The main value of ISO 31000 is to provide a structured basis for the integration of risk management processes and the establishment of a risk management culture. With regard to the use of ISO 31000 by mining companies, it should be noted that this international standard is generic in nature and, for its effective application, expert advice is required for its adaption and use.

Companies need to establish a transparent corporate standard for the identification, evaluation and management of TSF risks at each of their mine sites. In addition to the usual requirements of a corporate risk standard, the standard should also address the qualifications of the assessment team and require the implementation of critical control procedures. It should also provide guidelines to avoid complacency and conflicts of interest.

The position statement "Preventing catastrophic failure of tailings storage facilities" issued by the International Council on Mining and Metals, makes clear that enhanced efforts are required to ensure that:

> suitably qualified and experienced experts are involved in TSF risk identification and analysis, as well as in the development and review of effectiveness of the associated controls (ICMM 2016).

Regarding qualifications, the Rio Tinto management system standard states:

> Qualitative and quantitative risk analysis must be facilitated by competent personnel and include personnel with adequate knowledge and experience for the risk being evaluated (Rio Tinto 2015).

As already hinted earlier, we find the specification should be reworded: nowadays risk assessment is considered a discipline per se, which requires particular skills and knowledge. It is not a hobby … or something that can be "facilitated" by someone who has good knowledge of the hazards.

Furthermore, one of the most important factors is that the lead assessor be fully qualified to conduct such assessments. Expert knowledge in the design and operation of TSFs is actually not required, as the lead assessor is primarily required to lead the process, not be a factor in influencing the outcomes. With regard to the assessment team, it is important that a range of perspectives and experience be represented (Sect. 9.1.1) and include the participation of the EOR. What should not be allowed by the lead assessor is for persons to pull rank or dominate the discussions.

Incidentally, we are not keen on conducting risks assessments workshops in the way this is commonly done: unprepared people, no system definition (Chap. 8), no success/failure criteria definition (Sect. 7.7), etc. We always propose to perform preliminary definition and knowledge gathering actions and then enhance efficiency by leading small groups or even one-on-one meetings and interviews.

The ICMM position statement also remarked that performance criteria should be "… established for risk controls and their associated monitoring, internal reporting and verification activities" (ICMM website). ICMM further suggests that "Critical control management has been identified as an approach to managing low probability, high impact events such as catastrophic failures of tailings storage facilities". The identification of those issues that will require the highest level of attention is a necessary outcome of any risk assessment, and it cannot be the result of gut-feeling exercises.

Regarding risk management framework, leading practice would require that a company, working within the framework of ISO 31000, establish a corporate risk management standard that would include statements regarding the qualifications of audit assessment teams and require the identification of critical risks and their controls.

14.2   Manageable-Unmanageable and Strategic Risk Definition

Management is generally hard pressed to provide quick answers to questions about different aspects of the risks, such as which risks are:

- tolerable;
- intolerable but manageable, thus mitigation occurs by reducing the probability of failure;
- intolerable and unmanageable and hence require strategic shifts (altering the system).

These are obviously very different questions than the classic engineering questions: what FoS do we choose under various loading conditions? and in some cases, what deformations do we accept in which scenario/stage?

By using any of the explicit tolerance thresholds discussed above (Chap. 13) it is possible to provide a transparent definition of what constitutes a manageable risk: if a risk above tolerance (probability, consequence) can be brought under the selected tolerance threshold, before hitting the credibility limit of, say, 10^{-6}, by mitigative investments and risk transfer that still preserve the economic livelihood of a company, then that risk is manageable (yellow bubble in Fig. 14.1).

The key element here is a corporate/government choice of what level of mitigative investment preserves the economic livelihood of an entity. If the risk cannot be brought under the tolerance threshold as described, then it must be considered unmanageable. Unmanageable risks cannot be mitigated; they require strategic shifts

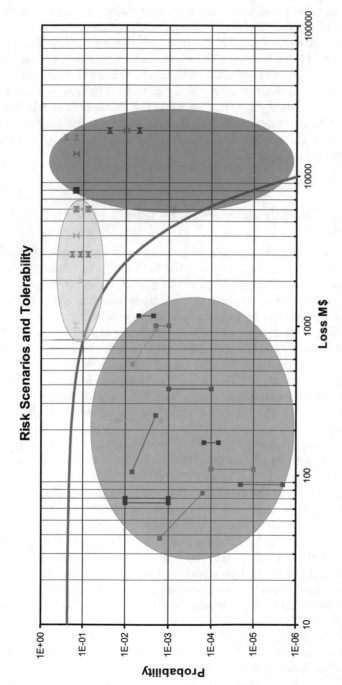

Fig. 14.1 Risk scenarios and tolerance. In blue: tolerable risks; in yellow: intolerable but manageable, in red: intolerable and unmanageable, hence strategic risks

in the corporation/government (red bubble in Fig. 14.1). Insurers and lenders may use this notion to select projects, clients, or create bundles of risks that solve the unmanageability of one or more specific projects in a portfolio.

Once risks (i.e., p, C) are evaluated for the entire hazard portfolio, then a graph like the one in Fig. 14.1 can be drawn. The orange curve is the risk tolerance (can be societal or corporate). We see three different classes of risk in Fig. 14.1, namely, the blue, yellow, and red. The blue are tolerable; therefore by definition they can be put aside until there are no more yellow and red risks. The yellow are the risks that can be mitigated by reducing their probabilities; they are therefore manageable. And the last class, the red, are the ones that cannot be mitigated unless we change the system. They are therefore unmanageable.

14.3   What to Do with Those Risks Families?

A risk assessment per se does not really help to make any decisions on risk reduction/accident prevention and other mitigative plans. It becomes rationally operational only when its results are compared with a threshold generally called "risk tolerance" (Fischhoff et al. 1982; Chap. 13). Risks that can be mitigated in a sustainable and economic way below tolerance by reducing their hazard probability are tactical risks. Instead, risks which require system's alterations (mitigations to reduce consequences and get the risk under tolerance) are strategic risks. In fact tactical risks fall under the responsibility of management, whereas strategic risks might require upper management to shift their objectives. To clarify this distinction with an example, we might say that buttressing a dam to reduce its breach probability is a tactical mitigation whereas changing the TSF location or not building a mine in a certain area are strategic mitigations.

The issue becomes more complicated when normalization of deviance is included in the discussion, as happened for the Oroville dam (https://www.riskope.com/2017/05/03/oroville-dam-risks-became-unmanageable/) (Sect. 3.1).

Figure 14.2 demonstrates the decrease of risk against increasing mitigative costs. When risks are very high, a relatively small investment generally makes it possible to reduce risks quickly whereas investments increase asymptotically when risks are reduced beyond a certain level. The graph shows the point at which an acceptable threshold of risk mitigation might be settled—stating explicitly that the mitigative costs will realistically be too high to achieve a theoretical total abatement of risk. The vertical line is set in the zone where risks are As Low as Reasonably Achievable (ALARA), As Low as Reasonably Practical (ALARP), or obey to the Best Available Control Technology (BACT) concept (Sect. 13.2.1).

Needless to say, the definition of these risk abatement levels is not common in many industries. In summary, the vertical line of Fig. 14.2 is depicted "in the commonly-accepted reasonable risk abatement zone", and its intersection with the risk abatement and mitigatory investment functions defines the residual risk and the investment necessary to attain it.

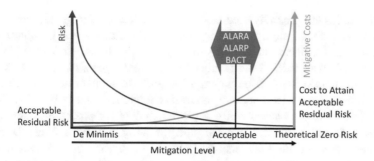

Fig. 14.2 Decrease of risk against increasing mitigative costs. The vertical line is set in the ALARP/ALARP/BACT zone. The position tends to slide toward the right, away from the optimum, due to public opinion pressure

14.4   Using the ORE Concept for Dams Systems

For our quantitative risk assessments we use our "universal" platform called ORE (Optimum Risk Estimates, ©Oboni Riskope Associates Inc., 2014–2019). Although ORE is a proprietary platform, we regularly teach its principles, which are summarised below. You are welcome to apply them "by hand" to simple cases, but beware: pretty soon any system goes beyond the "Excel worksheet" capabilities and properly structuring the risk register will become a challenge.

Ore follows a continuous loop as shown below in Fig. 14.3.

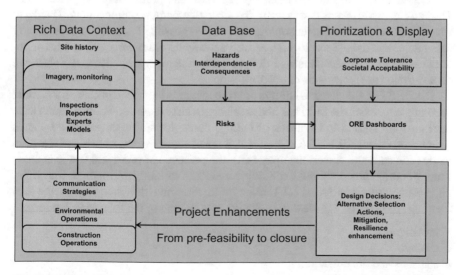

Fig. 14.3 The ORE (©Oboni Riskope Associates Inc.) continuous risk assessment/risk-based decision making workflow

ORE studies are generally used to support decisions (go/no-go, system changes, M&A, supply chain, etc.), tailings systems prioritizations, and insurance coverage decisions (business interruption, third-party liabilities, etc.) in mining and ancillary activities. ORE is the result of over twenty years of continuous development and applications. ORE makes it possible to include uncertainties, interdependencies, and societal and corporate risk tolerance, and is ISO 31000 compatible. ORE has been successfully deployed to date for processes, dams portfolios, linear facilities (roads, railroads, pipes), mining, etc. For tailings dams portfolios there is a specific ORE subset application (https://www.riskope.com/2018/11/21/tailings-dams-quantitative-risk-assessment/) named ORE2-Tailings (https://www.riskope.com/2018/05/02/ore2-tailings-dam-analysis/) that we routinely use to various degrees of sophistication.

14.5　Synergistic Methodologies

In order to fill the gaps and achieve the results outlined in Part I, the synergistic approach encompassing an updatable, scalable, drillable and convergent Quantitative Risk Assessment (QRA) platform and space or drone observation monitoring (Sect. 9.2.1) as main or complementary to extant classic monitoring programs (Sect. 9.1.2) are the instruments of choice.

Preliminary QRA deployment, using multiple data sources (Chap. 9), delivers initial estimates regarding probability of occurrence of various failure modes (Chap. 10), consequences of those failure modes (Chap. 12), and preliminary alert thresholds. They also provide results that assist in the setup of emergency procedures. Thanks to space or drone observation technologies, it is then possible to confirm and gradually calibrate extant data (Sect. 10.1.2), as well as validate old reports and their assumptions. In the next two sections we briefly describe the methodologies and tools as we deploy them for single dams or portfolio studies.

The ORE QRA devoted to tailings uses the methods described in Chaps. 9–12 and encompasses, when the SLM method is selected, thirty diagnostic points which have been custom tailored to the needs of tailings dams during the development phases (Oboni and Oboni 2013, 2016; Oboni et al. 2014). In fact, aspects such as construction mode (downstream, centerline, upstream), the presence of pipes and traffic at crest, water ponding, etc. were not explicitly considered in the SLM method. The thirty diagnostic points (see Sect. 11.2.1) are used to determine the "category" of the dam (ranging from 1—best in class to 4—poor), and belong to the following families:

- investigation;
- testing;
- analyses and documentation;
- construction;
- operations and Monitoring.

The category of a dam coupled with classic FoS (as defined by the engineers under various loadings) and the semi-empiric FoS/category curves leads to the annual probability of failure p_f estimate (Sect. 11.2). As there are various loading conditions and there may be pessimistic/optimistic evaluations of each one of the diagnostic nodes, uncertainties are transparently expressed. In order to allow swift, affordable analyses, the symptom-driven methodology anchored by hundred years of failure history and in-depth calibration has proven to offer the best solution when data are scarce and large portfolio require prioritization.

Table 14.1 summarizes the key space- or drone-observable elements considered when such observations become available. These, among others, contribute to the probability of failure and allow swift updates of the risks generated by a single dams or a portfolio of dams.

As per the consequences analyses, the considered space- or drone-observable data are:

- topography precise enough to allow, if needed, detailed calculations (the presence of man-made fills and ditches, even only a couple meters high/deep or a few meters wide, has significant impact on the flooding behaviour);
- bathymetry, if water bodies are present;
- land use/occupation, residences, infrastructure, lines, pipes, storage facilities, parking lots, etc.

Consequences can be evaluated following a number of methodologies and simulations, from the simplest to the most sophisticated, in function of the context of the study (Chap. 12).

In Fig. 10.5 (Sect. 10.2.1) we show the annual p_f results for a real-life tailings dams portfolio with 15 dams. We explained there that the horizontal axis shows the various dams of the dam portfolio as well as four bench-marking values, namely:

Table 14.1 Space observable diagnostic points summary

Elements	General diagnostic nodes
Dam system description	Tailings (beaches slopes, surface irregularities); water (size and position of pond with respect to dam crest, bathymetry, volume); reclaim pumps barge; equipment on crown (transport lines, spigotting, traffic); weir; intake tower/pennstock; diversion ditch
Construction	Material; berms and erosion; cross section; supervision; divergence from plans; known errors and omissions
Stability analyses	This data will come from client (if they have any records …), but possible instabilities (deformations, cracks, slumps) may be visible from space observation
Instability symptoms	Wet spots on the D/S face; streaming, ponding at toe (temperature differential); tailings deposited at toe
Settlements analyses	All this data will come from client (if they have any records …) but settlement may be visible from space observation

- Mount Polley and Samarco annual p_f evaluated with the ORE2_Tailngs methodology;
- the min-max values of the world-wide portfolio based on historic records (Oboni and Oboni 2013);
- the values obtained by a Ph.D. Thesis (Taguchi 2014) at UBC which attempted a theoretical estimate of the annual p_f of standard and dewatered tailings.

The vertical axis indicates the annual p_f. For each structure a yellow bar depicts the uncertainty related to the probability estimate. The portfolio bench-marking shows that in the considered case there are dams below the historic benchmark. Some overlap the benchmarks limits and some are above the upper limit, but are, however, still significantly lower than Mount Polley or Samarco estimates. We note the following:

- Additional studies and information would make it possible to narrow the uncertainties (length of the yellow bars).
- Mitigation on a specific dam would push the respective bar down.
- Long term lack of maintenance, climate change effects would tend to push the bars up.
- The combination of this graph with the cost functions for each dam gives the risks (Fig. 10.6). Interdependent dams can be analysed and their effect included in the portfolio analysis.

The development of a portfolio-specific risk tolerance threshold (Chap. 13) allows users to determine which risks actually really matter in a portfolio.

As a multidimensional consequence function is foreseen (consequences are generally multifaceted, see Sect. 7.6) it is possible to perform integrated comparison of project execution, community, legal, environmental, financial, technical and H&S risks (or whatever dimension the user may want to define). This convergent risk vision fosters healthy discussions and helps organizations to build consensus on decisions at the operational, tactical and strategic levels.

In Fig. 10.6 (in Sect. 10.2.2) various scenarios for "Dam 1" are displayed:

- Dam 1 mitigation shifts the risks downwards.
- Climate change effects increase the risks of the same structure.
- Lifts (also called raises in some cases) will also increase the risks, as losses will also increase.
- Finally inter-dependency between Dams 1 and 2 will generate the largest losses, but at a lower annual probability.

It is therefore possible to understand which are the most critical sources of threats to the tailings dam or compare each tailings dam's risks, e.g., which dam and hazard are loaded with the largest potential losses (split by type of loss: physical, BI, environmental, etc.) in a portfolio.

14.5.1 An Example of Results Display

As an example, we will look now at ORE results for a portfolio of 15 dams located in Canada (Oboni and Oboni 2017; Oboni et al. 2019), displayed in a different way.

Figure 14.4 is a probability-consequence graph where the 15 dams (aggregated risks per dam) have been plotted together with the corporate risk tolerance defined with the client and with Whitman's tolerance (Whitman 1984).

Figure 14.5 shows the same portfolio, but this time in terms of the intolerable aggregated risks for the dams above tolerance, and their relative percentage against the total intolerable risk of the portfolio.

It becomes immediately evident that two dams (9 and 12) represent together 99% of the intolerable risk of this portfolio, and (see Fig. 14.4) Dam 3 requires specific attention because it is corporately tolerable, but societally intolerable.

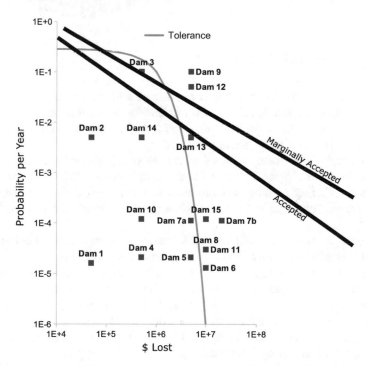

Fig. 14.4 Probability-consequence graph where the 15 dams are shown together with corporate risk tolerance and Whitman's societal thresholds (see Fig. 13.2)

Fig. 14.5 Same data of
Fig. 14.4, but very different
graphic display and
readability in terms of dams'
risk-prioritization and
mitigation roadmap

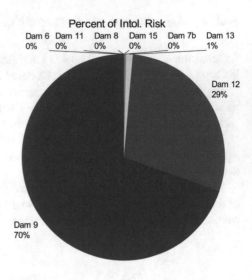

14.5.2   Using Risk Dashboards

Risk dashboards make it possible to communicate with management and the public. An example of possible selected instruments for such a dashboard is shown in Fig. 14.6a–d for the same portfolio depicted in Figs. 14.4 and 14.5. Figure 14.6a shows the dams sorted by decreasing total risk.

When corporate tolerance is added in the analyses (Fig. 14.6b) then it is possible to perform more sophisticated and useful prioritizations. It become possible to understand at first sight which dams represent an operational, tactical, i.e. intolerable but manageable, or strategic, i.e. intolerable and unmanageable risk (see Sect. 14.2; Fig. 14.1).

For example, as shown in Fig. 14.6c it is possible to show the tolerable (blue) and intolerable portion (orange) of risks for each dam in the portfolio.

Finally, as shown in Fig. 14.6d ranking of the dams by decreasing corporate intolerable risk can be performed. As we will show in Chap. 15, impacts on human life (societal tolerance, see Sect. 13.3.2) can be brought into the ranking and thus allow to deliver multi-objective optimization of risk mitigation road-maps.

This type of approach is scalable and drillable and can accommodate increasing level of information (during the corporate/project life). This is particularly interesting when initial (actuarial) data are scarce or non-existent and ample use of probabilities has to be made. ORE allows us to focus attention on the elements that generate the highest risks even in complex systems with hundreds or more components.

It becomes apparent that this type of approach makes it possible to compare alternatives and update "at any necessary" rate. It is the only risk management approach that can start static and then go all the way to real-time if the need exists and the resources are available.

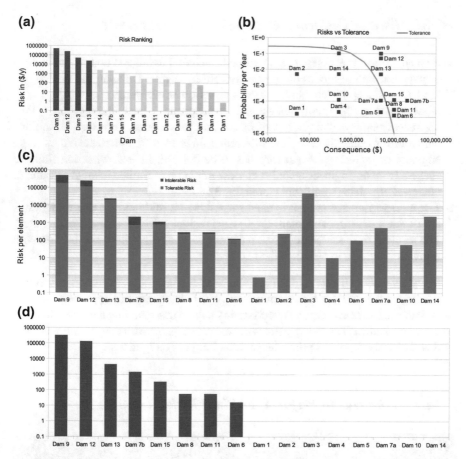

Fig. 14.6 ORE dashboard instrument example for the same portfolio (Figs. 14.4 and 14.5). **a** Total risk ranking for the dams in the portfolio. **b** The now well understood p–C graph with corporate tolerance. **c** Tolerable (blue) and intolerable portion (orange) of risks for each dam in the portfolio. **d** Ranking of the dams by decreasing corporate intolerable risk

14.6 How, Who, How Often?

In what follows we answer three basic questions related to residual risk assessment—"How, Who, How often"—and define for each the benefits/values brought in by the space- or drone-observation synergistic approach.

14.6.1   How to Perform a Residual Risk Assessment?

- First an a priori risk assessment is needed. It should detail and update evaluations of potential failure modes during the system life cycle. Defining the system is a fundamental step which requires lots of attention at inception.
- Understanding the multidimensional consequences and the system's failure/success criteria definition is of paramount importance. For example, tailings dam's failure oftentimes means different things to different stakeholders. e.g., engineer or regulators. A glossary has to be defined. Indeed, what constitutes a success from an engineering point of view might be of limited interest or value to other stakeholders (Oboni and Oboni 2018).
- The a priori risk assessment is used for RIDM on mitigation. Once mitigations are decided (and implemented) an a priori residual risk assessment is prepared. The residual risk assessment's risk register quantitatively integrates the data with mitigation leading to calculate the residual risks.

BENEFIT/VALUE: as scalable, drillable and convergent risk assessments are produced, no data will ever be lost or wasted. The risk register will be more detailed in areas that are better known, and uncertainties will be transparently conveyed in areas that are less known. The risk register will be ready to grow with the project/operation and will already support a priori decision making for mitigations.

14.6.2   Who Has to Perform Residual Risk Assessment?

> Summarizing the words of the ICMM (ICMM website) and UNEP (Roche et al. 2017): Reduce risk of dam failure by providing independent expert oversight; … monitoring and independent, third party review; … multiple, independent checks are required; … the adoption of regulations that require regular independent auditing; … independently reviewed and frequently updated risk assessments. Independence, updatability, risk assessments are key for the future.

The UNEP report cited at the very beginning of this book (Roche et al. 2017) also identifies a common practice that has to stop. The developer or design-engineers self-risk assessment has to stop as it is fraught with conflict of interest and inevitable biases. An independent risk assessor must become the new norm. The report identifies this requirement in distinct ways. For example, by stating:

> Establish independent waste review boards to conduct and publish independent technical reviews prior to, during construction or modification, and throughout the lifespan of tailings storage facilities.

This of course must include an independent risk assessment at every step. The report then adds:

- "Ensure any project assessment or expansion publishes all externalized costs, with an independent life-of-mine sustainability cost-benefit analysis." Including, of course the risks.
- "Require detailed and ongoing evaluations of potential failure modes, residual risks and perpetual management costs of tailings storage facilities."
- "Reduce risk of dam failure by providing independent expert oversight" done by independent risk assessor to maintain good and unbiased oversight. This will "Ensure best practice in tailings management, monitoring and rehabilitation" (Roche et al. 2017).

BENEFIT/VALUE: The independent risk assessor will ensure a drastic reduction of conflict of interest and the delivery of unbiased risk reports. Space or drone Observation and a transparent risk platform as described earlier will deliver to the independent risk assessor unbiased data interpreted using auditable rules, transparent risk registers. All requirements of UNEP will be met.

14.6.3 How Often Should a Residual Risk Assessment Be Performed?

Ideally, a residual risk assessment should be performed every time the conditions change. Conditions can range from weather patterns to managerial changes. Of course, any addition/alteration to the overall system should also trigger an update. We have experienced that situations oftentimes change quickly and adaptive changes may require gathering different types of information from different areas. Forms of traditional (risk matrix: FMEA, PIGs) risk assessment update are too fuzzy and lack definition to identify anomalies and allow for swift adaptations.

In theory, updates could occur in "real time" if IoT or other high frequency observations methods are used. Practice and rate of movements, evolution, etc. dictate the optimum updates frequency.

BENEFIT/VALUE: Great economy of personnel, traditional instrumentation, time. Reduced risks for personnel on the ground in hazardous areas. It is possible to change frequency of monitoring to adapt to new and emerging situations.

14.7 Conclusions

The various examples discussed above show the benefits found in linking multi-temporal objective space or drone observation with a dynamic convergent QRA platform in mining projects and operations. The link developed to date enables us to "measure" and give a sense to a complex problem. It makes it possible to:

- transparently compare alternatives;
- discuss rationally and openly the survival conditions; or

• evaluate the premature failure of a structure.

The link between the context of space or drone observation "rich data" and the risk assessment platform uses Bayesian updates of probabilities, frequencies and other selected parameters to distill the data used in the risk assessment. Connecting a dynamic quantitative risk analysis platform with a high-performance data gathering technique reduces costs, avoids blunders, and constitutes a healthy management practice, especially for long-term projects requiring short- or long-term monitoring, including, of course, site restorations.

References

[AG TMH 2016] Australian Government (2016) Tailings Management: Leading Practice Sustainable Development Program for the Mining Industry. https://www.industry.gov.au/sites/g/files/net3906/f/July%202018/document/pdf/tailings-management.pdf

[ANCOLD 2003] Australian National Committee on Large Dams (2003) Guidelines on risk assessment. Sydney: ANCOLD

Barker M (2011) Australian risk approach for assessment of dams. In: The 21st century dam design — advances & adaptations. Proceedings of the 31st Annual USSD Conference, San Diego, CA. Denver, US: Society on Dams

Brown ET (2017) Reducing risks in the investigation, design and construction of large concrete dams Journal of Rock Mechanics and Geotechnical Engineering 9(2): 197–209

Fischhoff B, Lichenstein S, Slovic P, Derby SC, Keeney R (1982) Acceptable risk. Cambridge MA, Cambridge University Press

[ICMM Review 2016] International Council on Mining & Metals (2016) The International Council on Mining and Metals, 2016 Annual Review, Enhancing Mining's contribution to society https://www.icmm.com/website/publications/pdfs/annual-review/2016_icmm_annual-review

[ICOLD 2013] International Commission on Large Dams (2013). Sustainable design and post-closure performance of tailings dams, Bulletin 153, ICOLD, Paris

Oboni C, Oboni F (2013) Factual and Foreseeable Reliability of Tailings Dams and Nuclear Reactors -a Societal Acceptability Perspective, Tailings and Mine Waste 2013, Banff, AB, November 6 to 9, 2013

Oboni C, Oboni F (2018) Geoethical consensus building through independent risk assessments. Resources for Future Generations 2018 (RFG2018), Vancouver BC, June 16–21, 2018

Oboni F, Oboni C (2016) A systemic look at tailings dams failure process, Tailings and Mine Waste 2016, Keystone CO, USA, October 2–5, 2016 https://www.riskope.com/wp-content/uploads/2016/10/Paper-17_A-systemic-look-at-TD-failure-TMW-2016_07-11_revised.pdf

Oboni F, Oboni C (2017) Screening Level Risk Assessment for a Portfolio of Tailings Dams, Canadian Dam Association (CDA, ACB) Bulletin 28(4): 17–29

[Oboni et al. 2014] Oboni F, Oboni C, Caldwell J (2014) Risk assessment of the long-term performance of closed tailings, Tailings and Mine Waste 2014, Keystone CO, USA, October 5–8, 2014

[Oboni et al. 2019] Oboni F, Oboni C, Morin R (2019) Innovation in Dams Screening Level Risk Assessment, International Commission on Large Dams/Commission Internationale des Grands Barrages (ICOLD/CIGB), Ottawa, Canada, June 9–14, 2019

[Oboni et al. 2013] Oboni F, Oboni C, Zabolotniuk S (2013) Can We Stop Misrepresenting Reality to the Public?, CIM 2013, Toronto. https://www.riskope.com/wp-content/uploads/Can-We-Stop-Misrepresenting-Reality-to-the-Public.pdf

[Rio Tinto 2015] Rio Tinto Management System (2015) Standard https://www.riotinto.com/documents/RT_Management_System_Standard_2015.pdf

Roche C, Thygesen K, Baker, E (eds.) (2017) Mine Tailings Storage: Safety Is No Accident. A UNEP Rapid Response Assessment. United Nations Environment Programme and GRID-Arendal, Nairobi and Arendal, ISBN: 978-82-7701-170-7 http://www.grida.no/publications/383

Taguchi G (2014) Fault Tree Analysis of Slurry and Dewatered Tailings Management – A Frame Work, Master's thesis, University of British of Columbia

Whitman RV (1984) Evaluating calculated risk in geotechnical engineering. J. Geot. Engineering 110(2): 145–188

Part III
Case Histories and a Look into the Future

Part III, Case Histories and a Look into the Future, features some examples of deployments and concludes with a look into what we expect might happen in the years and decades to come.

Chapter 15
Risk-Informed Decision Making

This chapter, Risk-Informed Decision Making, presents a set of case histories used as deployment examples. This chapter is devoted to the relative quantitative risk assessment of a portfolio of tailings dams. It aims to establish relative probabilities and consequences, allowing comparisons and prioritization of the portfolio's risks but makes no attempt to deliver absolute probabilities for one specific object. This chapter shows how the results of a modern QRA lead to the design of a mitigative roadmap which covers corporate, societal, and legal interests.

Now that we have identified an ISO-31000-compliant, convergent, scalable, drillable, updatable risk management system (Chap. 14) based on a clearly defined glossary (Sect. 5.2), it is time to show examples of risk-informed decision making (RIDM) applied, for example, to a portfolio of dams.

In real life this type of assessment obeys a number of rules:

- It does not include checking the extant files for either geomechanical analyses/parameters selection or the absence of errors and omissions. However, it does include checking to see whether the necessary data were reportedly collected and worked through the design in a coherent way.
- It does not replace any work performed by the project engineers, engineers of record, inspectors, monitoring experts and other professionals involved with the structure.

  - It interprets information delivered in the geostructure's files using a variety of reproducible and transparent procedures described in earlier chapters (Chaps. 10 and 11) with the objective of establishing relative probabilities, allowing comparisons and prioritization of risks.
  - It examines equipment on the geostructures' crown and the reservoir's level control (if applicable) to include their influence on the geostructure's probability of failure.
  - It does not constitute an attempt to deliver absolute probabilities for one specific object.

© Springer Nature Switzerland AG 2020
F. Oboni and C. Oboni, *Tailings Dam Management for the Twenty-First Century*,
https://doi.org/10.1007/978-3-030-19447-5_15

– It includes consequences, when applicable, analyses from extant reports and from recent research (Chap. 12).

• It is useful to remember that historically the probability of dam failure ranges from 10^{-3} to $2 * 10^{-4}$ for "generic" tailings dams (see Table 4.1, Sect. 4.1), with respect to a range from 10^{-5} to $3 * 10^{-6}$ for top of class water dams, which can be considered at "credibility threshold". These ranges can be used to benchmark the result of this assessment to historic "world-wide portfolio" performances (Chap. 4). As noted earlier, these historical data have to be taken with prudence, due to the inconsistent way of attributing the principal cause of failure and, for example, the large number of "unknown", "not available" and "other" categories.

We will use a portfolio of four dams, namely Dam 1, 2, 3, 4, with one major external interdependency. One of the four dams will be studied at two distinct raise levels (Dam 1A, Dam 1B).

These dams are all real-life dams belonging to different clients around the world. In order to maintain confidentiality, we have altered the dams profiles and the geotechnical parameters of the materials, and we have performed several data alterations.

Due to obvious limitations of space we will develop each case only to the level required for the discussion and the points we want to make.

15.1   Portfolio Description

This section describes the system constituted by the dams and their internal and external interdependencies. In all the dams the engineers declared the bedrock to be competent and free of weak layers and unfavorable discontinuities. Further, lique-faction (see Sect. 9.3) was disregarded by the engineers as a potential failure factor.

15.1.1   General Descriptions of the Dams

Dam 1A is an intermediate situation of Dam 1, i.e., the raise of Dam 1 at 100 m altitude (Fig. 15.1).

Piezometers show phreatic levels at bedrock in zones 1 and 2 downstream of zone 3.

Inclinometers have not shown any significant discontinuity, i.e., there is no suspicion of instability. Geotechnical parameters of the various soils are defined in Table 15.1.

Dam 1B (Fig. 15.2) is the same dam as Fig. 15.1, but at its final altitude of 162 m. No changes of conditions with respect to Fig. 15.1.

Dam 2 is a water retention dam (Fig. 15.3) which displays an inter-dependency with Dam 3.

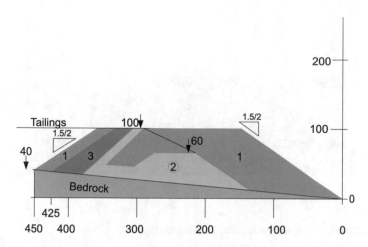

Fig. 15.1 Dam 1A cross section: raise at altitude 100 of Dam 1

Table 15.1 Geotechnical parameters of the soils in cross sections of Dams 1A and 1B

Zone	Name	γ (kN m^{-3})	C′ (kPa)	φ' (°)
1	Rockfill 1	24	0	38
2	Rockfill 2	22	0	35
3	Low perm. material	21	18	31.5
4	Compacted material	22	10	28

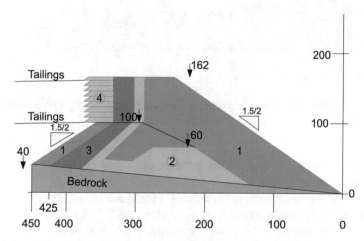

Fig. 15.2 Dam 1B or final stage, height 162 m, of Dam 1, centerline rockfill dam

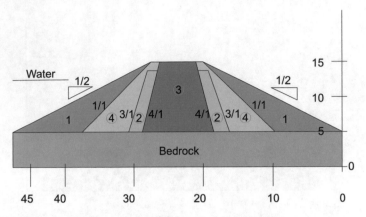

Fig. 15.3 Dam 2 centerline cross section built in one raise, height 15 m

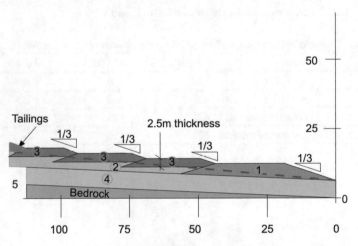

Fig. 15.4 Dam 3 upstream berms on starter berm cross section, total height 30 m at the moment of the analysis

Piezometric surface at bedrock downstream of dam, at water level upstream. No movements detected to date.

Dam 3 is an upstream structure (Fig. 15.4) constituted by a compacted till starter berm on which upstream compacted waste rock berms have been built.

The piezometric level follows dotted blue line. Inclinometers have shown 15 mm deformation within layer 4, between bedrock (stable) and surface.

Finally, Dam 4 was introduced in Sect. 11.1.2. It is an upstream dam of obsolete design similar to recently failed dams. Details were shown in the referenced section, and drawings are not available for confidentiality reasons.

15.1.2 Interdependencies of the Dams

There is an external interdependency between Dam 2 and Dam 3 insofar as a failure of Dam 2 would provoke an erosion at the toe of Dam 3, as we will describe later.

Furthermore, Dam 3 has internal interdependencies due to possible local failures of raises.

15.1.3 Ancillary Structures and Other Dam Conditions

There are no penstocks systems in the portfolio's dams, except for Dam 4.

There are tailings pipelines at the crest of Dams 1A, 1B and 3 while Dam 4 is in the process of deactivation, thus there is no active spigotting at its crest.

There are vehicular and subcontractor traffic at the crest of all dams.

15.1.4 Scheme of the System

The "map" of the system is schematically displayed in Fig. 15.5. The map helps to define the boundaries of the dam systems and the overarching portfolio (see Sect. 8.2.2). However, the map is a symbolic one, with no scale. Tailings are depicted in grey; the water of pond 2 and watercourses in blue; valleys bottoms are white; and major residential and industrial structures are displayed as orange rectangles. Roads appear as single black lines, whereas a railroad, in a dotted line, crosses the map from east to west in the south of the map.

The map has the objective of allowing a preliminary estimate of potential consequences by looking at the land use downstream of each dam. The level of details is variable: the more potential targets of high value, the higher the detail in order to enable consequence evaluations. It is very important to focus on the consequences, as engineering analyses oftentimes focus on the hazard side of the risk equation and neglect the consequence side (Eq. 1.1).

15.1.5 Geotechnical Conditions as Reported by Engineering Reports

The geotechnical materials for Dams 1A and 1B are defined in Table 15.1.

The geotechnical materials for Dam 2 are defined in Table 15.2.

The geotechnical materials for Dam 3 are defined in Table 15.3.

Fig. 15.5 "Map" of the system of dams constituting the portfolio

Table 15.2 Geotechnical parameters of the soils in cross sections of Dam 2

Zone	Name	γ (kN m^{-3})	C' (kPa)	φ' (°)
1	Compacted till	18	5	30
2	Sandfill	20	0	30
3	Clay	18	53–37	0
4	Compacted material	20	0	38

Table 15.3 Geotechnical parameters of the soils in cross sections of Dam 3

Zone	Name	γ (kN m^{-3})	C' (kPa)	φ' (°)
1	Compacted till	21	0	35
2	Tailings	19.5	0	30–(5.7) (liqu.)
3	Crushed waste	20	0	38
4	Silty clay	16.5	0	28

Geotechnical conditions for Dam 4 are scarce and have been summarized in Sect. 11.1.2.

15.2 Engineering Stability Analyses of the Dams

The engineers have defined the following FoS for the various dams under various analysis conditions:

- USA undrained stress;
- ESA effective stress;
- pseudostatic;

as shown in Table 15.4.

Let's comment on the values of the FoS of the dams in this portfolio.

An exception being made of Dam 4, which is clearly a "bad apple", the FoS have very similar values in USA for Dams 2 and 3, in ESA for Dams 1A, 1B and 2. Finally, there are very similar FoS under pseudo-static conditions for Dams 1, 2 and 3. We will see later that the image we obtain with the probability of failure is much more defined, with more evident variations from one dam to the next. This extra definition is very important when a risk prioritization of a large portfolio is sought.

Table 15.4 Results of the engineers' stability analyses for the various dams of the portfolio

Dam	USA	ESA	Pseudo-static
1A	n/a	1.61–1.68	1.18–1.23 for 1/475 Operating base earthquake (OBE) and 1/1500 MCE
1B	n/a	1.59–1.65	1.15.1.26 for 1/475 (OBE)
2	Rapid drawdown 1.63–1.76	1.59–1.68	1.28–1.33 for 1/1000
3	1.6	1.29–1.35	1.2 for 1/2475 maximum credible earthquake (MCE)
4	Refer to Tables 11.1 and 11.2 in Sect. 11.1.2: FoS = 1 to 1.35, average 1.17		

This will avoid our falling in the trap of characterising as "safe" dams that have a probability of failure well in the range of credibility.

15.3   Conditions from Inception to the Date of Analysis of the Dams (Physical and Governance)

In this section we summarize the conditions over the known history of the dams, as has resulted from the reading of extant reports and records. Such a summary could also result from interviews and other classic means (Sect. 9.1). Should modern hazard identification and monitoring data (Sect. 9.2) such as space or drone observation history be available, then those results should also be integrated in the dam descriptions. IoT and other data analyses would also complement this information.

Of course, all the data and interpretations contained herein should be stored (and retrievable by drills), updated in a portfolio data base (Hazard and Risk Register).

It is obvious that the necessary complexity of the structure of the register records requires a carefully designed software. This is the reason why we designed the ORE2_Tailings application (Sect. 14.4).

15.3.1   Dam 1A and Dam 1B

Construction:

- As described above, this is a centerline-rockfill tailings dam analyzed at two stages (100, 162 m).
- Non-erodable by assumed extreme meteorological conditions.
- Pipeline and traffic at crest.
- Supervision of the dam is excellent in terms of both frequency and the of skills of the supervisor.
- No geometric divergences have been noted between the plans and the actual structures and minor intermittent seepage is monitored at one location at the toe.
- There are no known errors or omissions in this project, which has been third-party reviewed by competent independent reviewers.

Geotechnical investigations and testing:

- Boreholes along the dam layout were regular. However, at less than 1/100 m, they were rather short, barely entering the bedrock.
- No continuous sampling was performed.
- Vane tests and cone tests were numerous but poorly distributed due to limitations on access.
- Soil classification tests were performed in a reasonable number and at a reasonable frequency,

- Geomechanical tests were numerous and distributed.
- A significant number of residual strength, oedometer and triaxial tests were performed.

Analyses and documentation:

- The project is deemed to be of good quality by a competent engineering firm and a skilled team.
- The "as built" plans display no significant imperfections and variations.
- The alteration plans have also been regularly updated.

Stability analyses:

- As shown in Table 15.4, effective and pseudostatic stability analyses were performed by the engineers.
- Settlement analyses were performed.
- Liquefaction was dismissed by the engineers and we agreed.
- Internal erosion analysis was estimated by engineers as not being a problem.

Operations and monitoring:

- Pore pressure is measured with numerous vibrating wire piezometers.
- Deformations are monitored by topographic observations and inclinometers.
- There is an active Independent Geotechnical Review Board (IGRB).
- There is an annual inspection run by a reputable third party.

Maintenance and repairs:

- Repairs are carried out in a timely manner when damages are identified during the inspection.

15.3.2 Dam 2

Construction:

- As described above, this is a water retention centerline dam built in one raise with selected materials.
- Non-erodable by assumed extreme meteorological conditions.
- No pipeline at crest.
- Supervision of the dam was mediocre in terms of both frequency and the skills of the supervisor.
- Some locally significant geometric divergences have been noted between the plans and the actual structures and minor intermittent seepage is monitored at one location at the toe.
- There are no known errors or omissions in this project, which has been third-party reviewed by competent independent reviewers.

Geotechnical Investigations and testing:

- There are only a few boreholes along the dam layout (fewer than 1/100 m); they were at least as deep as the structure is high.
- No continuous sampling was performed, but fifteen trenches were dug, although poorly distributed along and across the layout.
- Vane tests and cone tests were numerous but poorly distributed due to limitations of access.
- Soil classification tests were performed in a reasonable number and at a reasonable frequency.
- Geomechanical tests were limited and poorly distributed.
- A very limited number of residual strength, oedometer and triaxial tests were performed.

Analyses and documentation:

- The project is deemed to be of good quality by a competent engineering firm and a skilled team.
- The "as built" plans display some significant imperfections and variations.
- the alteration plans also have some significant approximations.

Stability analyses:

- As shown in Table 15.4, effective drawdown and pseudostatic stability analyses were performed by the engineers.
- No settlement analyses were performed.
- Liquefaction was dismissed by the engineers and we agreed.
- Internal erosion analysis was estimated by engineers as not being a problem.

Operations and monitoring:

- Pore pressure is measured with six vibrating wire piezometers.
- Deformations are monitored by topographic observations.
- There is no active Independent Geotechnical Review Board (IGRB).
- There is an annual inspection ran by a reputable third party.

Maintenance and repairs:

- No repairs are carried out when damages are identified during the inspection.

15.3.3   Dam 3

Construction:

- As described above, the raises are upstream, made of compacted crushed rock, on tailings.

- Engineering studies have determined the berms (raises) are non-erodable by extreme meteorological conditions or even by a breach of the tailings pipeline.
- Supervision of the dam is of good quality in terms of both frequency and the skills of the supervisor.
- Minor geometric divergences have been noted between the plans and the actual structures and minor intermittent seepage is monitored at one location at the toe of the starter berm.
- There are no known errors or omissions in this project, which has been third-party reviewed by competent independent reviewers.

Geotechnical Investigations and testing:

- There are only a few boreholes along the dam layout (fewer than 1/100 m); they are generally short (the engineers argue the rock is of a sound nature and no weaknesses are expected), barely penetrating into the assumed bedrock.
- No continuous sampling was performed.
- Penetrometers, vane tests and cone tests were numerous but poorly distributed due to limitations of access.
- Soil classification tests were performed in a reasonable number and at a reasonable frequency.
- Geomechanical tests were numerous but poorly distributed.
- A limited number of residual strength, oedometer and triaxial tests were performed.

Analyses and documentation:

- The project is deemed to be of good quality by a competent engineering firm and a skilled team.
- The "as built" plans display some imperfections and variations.
- The alteration plans have also some approximations.

Stability analyses:

- As shown in Table 15.4, effective, undrained and pseudostatic stability analyses were performed by the engineers.
- No settlement analyses were performed.
- Liquefaction was dismissed by the engineers. Without proof we consider this as a likely event due to dam's construction and the material underlying the raises.
- Internal erosion analysis was estimated by engineers as not being a problem. We consider again this as a likelihood, given surface observations.

Operations and monitoring:

- Pore pressure is measured with over twenty vibrating wire piezometers.
- Four inclinometers are regularly read (although they are too short in our opinion).
- There is an active and competent Independent Geotechnical Review Board (IGRB).
- There is an annual inspection run by a reputable third party.

Maintenance and repairs:

- Some repairs are carried out when damages are identified during the inspection and after the IGRB discusses them.

15.3.4   Dam 4

In Sect. 11.1.2 the dam situation was qualified as follows for the north section of the dam (see Tables 11.1 and 11.2):

- Slopes are steeper than designed;
- Failure is incipient toward the Eastern abutment;
- There is a monitoring system, but there is limited information regarding the readings, frequency, etc.
- There is uncontrolled erosion of the downstream (D/S) slopes.

15.4   Success and Failure Criteria

As we explained in Sect. 7.7, a risk assessment should always start with the definition of the success and failure criteria. Indeed, the failure metric needs to be clearly stated in order to avoid plunging into a state of confusion from the first contact with the public (see Sect. 15.11.1). For example, in a dam risk assessment, if the dam "survives" a given hazard but the water source for the local population is contaminated, it should be considered a failure. Too often we have seen a failure considered only when a dam is breached, or looked at only from a stability point of view, disregarding other issues, and forgetting the system's success metric. This became obvious with the Oroville hydrodam's forensic analysis (see Sect. 3.1), which stated:

> The Independent Forensic Team reportedly formed the impression that most Department of Water Resources—DWR—staff and those involved in the Potential Failure Modes Analysis—PFMA—studies considered the use of the emergency spillway in terms of only an "extreme" flood event. The Independent Forensic Team notes that a "1 in 100" year storm might be considered an "extreme" event in an operational sense. However, from a dam safety viewpoint an "extreme" dam safety would be a much larger storm.

Incidentally, the Oroville case was again an example of unclear glossary and definitions leading to misleading evaluations of risks. As the Oroville report states, failure modes can be excluded from detailed consideration simply due to the definition of "failure" that is used. In our practice we always start a risk assessment by defining the system, its success and failure criteria, and multidimensional consequence metrics (Oboni and Oboni 2018).

For this case study we will use a success criterion as follows: A dam is successful if it does not breach due to "birth defects", construction, operations, management and monitoring malfunctions during its expected life. Slow contaminant releases, slow/small entity outflows are excluded from this analysis. A truly holistic view should of course include slow releases, slow/small entity outflows into the environment. However, their inclusion would add complexity to the case study without adding insightful information on the procedures to be implemented.

15.5 Probability of Failure of the Portfolio's Dams

In this section we will use a combination of techniques as described in Chaps. 10 and 11, based on availability of data and ease of implementation. As the risk register of the portfolio should be built "for future evolution", scalability should be insured. That means that if data for a given dam receive a complement, then those records could be easily updated either using Bayesian concepts (see Sect. 10.1.2) or simply new data inputs.

When the SLM method (Silva et al. 2008), later modified by Altarejos-García (Altarejos-García ct al. 2015) is used, Table 11.4 adapted for each portfolio's dam is displayed in the next sections with explicit indication of the category determination.

The thirty diagnostic "points" used in ORE2_Tailings assessments of tailings dams (see Sects. 11.2.1, 11.2.3 and 14.5) are summarized as follows:

- Physical aspects of the dam and its equipment (weirs, pipes, spigots, penstocks, ctc.);
- Construction: type of materials, cross section, supervision, berms and erosion, divergence from plans, etc.;
- Geotechnical investigations and testing;
- Prior analyses and documentation of the project;
- Various stability, deformation, erosion, liquefaction aspects:

  - stability analyses (of various types, ESA, USA, pseudo-static, etc.)
  - instability symptoms
  - settlements (actual and analyses)
  - liquefaction
  - internal erosion

- Operations, monitoring, maintenance and repairs.

Based on the list above, evaluations can be easily updated to take into account generally available or observable data. Furthermore, the lack of knowledge—i.e., uncertainties on data or missing data—is explicitly entered in the evaluations. Explicit consideration of uncertainties is a fundamental step towards reasonable, transparent and ethical risk assessments.

15.5.1   Dam 1A, 1B

Table 15.5 displays the SLM results for Dam 1A, 1B.

Table 15.5   SLM (Silva et al. 2008) results for Dams 1A and 1B

Cat.	Design			Build	Operate	
	D1	D2	D3	CO	OM	
	Investigation	Testing	Analyses and documentation	Construction	Operations and monitoring	Category
I Best	Evaluate design and performance of nearby structures	Run lab tests on undisturbed specimen at field conditions	Determine FS using effective stress parameters based on measured data for site	Full time supervision by qualified engineer	Complete performance program including comparisons between predicted and measured	1.4 to 1.6
	Analyze historic aerial photos	Run strength tests along field effective total and effective stress paths	Consider field stress path in stability determination	Construction control tests by qualified engineers and technicians	No malfunctions (cracks slides etc.)	
	Locate all non uniformities	Run index field tests (field vane, come penetrometer) to detect anomalies	Prepare flow nets for instrumented sections	No errors or omissions	Continuous maintenance	
	Determine site geologic history	Calibrate equipment and sensors prior to tests	Predict pore pressure and other relevant performance parameters for instrumented sections	Construction report clearly documents construction activities		
	Determine subsoil profiles using continuous sampling		Have design report clearly document parameters and analyses used for design			
	Obtain undisturbed samples for testing		No errors or omissions			
	Determine field pore pressure		Peer review			

(continued)

Table 15.5 (continued)

	0.2–0.3	0.2–0.3	0.2	0	0	
	Evaluate design and performances of nearby structures	Run standard tests on undisturbed specimen	Determine FS using effective stress parameters and pore pressure	Part time supervision by qualified engineer	Periodic inspection by qualified engineer	
	Exploration program tailored to project conditions by qualified engineer	Measure pore pressure in strength tests	Adjust for significant differences between field stress paths and stress path implied in analysis that could affect design	No errors or omissions	No uncorrected malfunctions	
		Evaluate differences between laboratory tests conditions and field conditions			Selected field measurements	
II Above Average					Routine maintenance	
	0.2 to 0.3	0.2 to 0.3	0.2	0.4	0.4	
	Evaluate performances of nearby structures	Index tests on samples from site	Rational analyses using parameters inferred from index tests	Informal construction supervision	Annual inspection by qualified engineer	
III Average	Estimate subsoil profile from existing data and borings				No filed measurements	
					Maintenance limited to emergency repairs	
	0	0	0	0	0	
	No field investigation	No laboratory tests on samples obtained at the site	Approximate analyses using assumed parameters	No construction supervision by qualified engineer	Occasional inspection	
				No construction control tests	No field measurements	
IV Poor	0	0	0	0	0	

The SLM methods gives Dams 1A and 1B category 1.4–1.6 based on the selections made in Table 15.5. It is worth noting that the SLM method does not contain any provision for water balance or freeboard, as it is a general slope methodology, while ORE2_Tailings includes those factors and many other dam-specific ones. ORE2_Tailings delivers Category 1.45–2.22, mainly because of the short, relatively widely-spaced boreholes performed during the investigations. The rockfill nature of the downstream slope of the dams makes it insensitive to erosion, as pointed out by the engineers, and to a possible crest pipeline breach.

15.5.2   Dam 2

Table 15.6 displays the SLM results from Dam 2.

As shown in Table 15.6, the SLM method yields category 2.2–2.4. In the case of Dam 2, where there are little tailings-dam-specific features, the correspondence of SLM with ORE2_Tailings Category (2.17–2.36) is very good.

15.5.3   Dam 3

Table 15.7 displays the SLM results from Dam 3.

The SLM method delivers category 1.6–1.7 (Table 15.7). Because the SLM method does not consider liquefaction, the divergence here is again significant: ORE2_Tailings yields category 1.83–2.37.

Dam 3 could be significantly eroded were the crest pipeline to breach and remain unobserved for more than 12 h.

15.5.4   Dam 4

Table 15.8 displays the SLM results from Dam 4.

The ORE2_Tailings category for such a dam, characterized by the paucity of data and ignorance on many aspects of its history, is Category 3.3–3.8.

The SLM method delivers Category 3–3.2 again because it does not look at specific tailings dams deficiencies. Dam 4 is being deactivated, so there are no active pipelines at its crest.

Table 15.6   SLM (Silva et al. 2008) results for Dam 2

Cat.	Design			Build	Operate	
	D1	D2	D3	CO	OM	
	Investigation	Testing	Analyses and Documentation	Construction	Operations and monitoring	Category
	Evaluate design and performance of nearby structures	Run lab tests on undisturbed specimen at field conditions	Determine FS using effective stress parameters based on measured data for site	Full time supervision by qualified engineer	Complete performance program including comparisons between predicted and measured	
	Analyze historic aerial photos	Run strength tests along field effective total and effective stress paths	Consider field stress path in stability determination	Construction control tests by qualified engineers and technicians	no malfunctions (cracks slides etc.)	
	Locate all non uniformities	Run index field tests (field vane, come penetrometer) to detect anomalies	Prepare flow nets for instrumented sections	No errors or omissions	Continuous maintenance	
	Determine site geologic history	Calibrate equipment and sensors prior to tests	Predict pore pressure and other relevant performance parameters for instrumented sections	Construction report clearly documents construction activities		
	Determine subsoil profiles using continuous sampling		Have design report clearly document parameters and analyses used for design			
	Obtain undisturbed samples for testing		No errors or omissions			
I Best	Determine field pore pressure		Peer review			

(continued)

Table 15.6 (continued)

	0	0.2–0.3	0.2	0	0	
II Above Average	Evaluate design and performances of nearby structures	Run standard tests on undisturbed specimen	Determine FS using effective stress parameters and pore pressure	Part time supervision by qualified engineer	Periodic inspection by qualified engineer	2.2 to 2.4
	Exploration program tailored to project conditions by qualified engineer	Measure pore pressure in strength tests	Adjust for significant differences between field stress paths and stress path implied in analysis that could affect design	No errors or omissions	No uncorrected malfunctions	
		Evaluate differences between laboratory tests conditions and field conditions			Selected field measurements	
					Routine maintenance	
	0	0	0	0	0	
III Average	Evaluate performances of nearby structures	Index tests on samples from site	Rational analyses using parameters inferred from index tests	Informal construction supervision	Annual inspection by qualified engineer	
	Estimate subsoil profile from existing data and borings				No filed measurements	
					Maintenance limited to emergency repairs	
	0.6 to 0.7	0.2 to 0.3	0.2	0.6	0.6	
IV Poor	No field investigation	No laboratory tests on samples obtained at the site	Approximate analyses using assumed parameters	No construction supervision by qualified engineer	Occasional inspection	
				No construction control tests	No field measurements	
	0	0	0	0	0	

Table 15.7   SLM (Silva et al. 2008) results for Dam 3

Cat.	Design			Build	Operate	
	D1	D2	D3	CO	OM	
	Investigation	Testing	Analyses and documentation	Construction	Operations and monitoring	Category
	Evaluate design and performance of nearby structures	Run lab tests on undisturbed specimen at field conditions	Determine FS using effective stress parameters based on measured data for site	Full time supervision by qualified engineer	Complete performance program including comparisons between predicted and measured	
	Analyze historic aerial photos	Run strength tests along field effective total and effective stress paths	Consider field stress path in stability determination	Constructi on control tests by qualified engineers and technicians	No malfunctions (cracks slides etc.)	
	Locate all non uniformities	Run index field tests (field vane, come penetrometer to detect anomalies	Prepare flow nets for instrumented sections	No errors or omissions	Continuous maintenance	
I Best	Determine site geologic history	Calibrate equipment and sensors prior to tests	Predict pore pressure and other relevant performance parameters for instrumented sections	Constructi on report clearly documents construction activities		1.6 to 1.7
	Determine subsoil profiles using continuous sampling		Have design report clearly document parameters and analyses used for design			

(continued)

Table 15.7   (continued)

	Obtain undisturbed samples for testing		No errors or omissions		
	Determine field pore pressure		Peer review		
	0	0.2	0	0.2	0
II Above Average	Evaluate design and performances of nearby structures	Run standard tests on undisturbed specimen	Determine FS using effective stress parameters and pore pressure	Part time supervision by qualified engineer	Periodic inspection by qualified engineer
	Exploration program tailored to project conditions by qualified engineer	Measure pore pressure in strength tests	Adjust for significant differences between field stress paths and stress path implied in analysis that could affect design	No errors or omissions	No uncorrected malfunctions
		Evaluate differences between laboratory tests conditions and field conditions			Selected field measurements
					Routine maintenance
	0.4 to 0.5	0.2	0.4	0.2	0.4
	Evaluate performances of nearby structures	Index tests on samples from site	Rational analyses using parameters inferred from index tests	Informal construction supervision	Annual inspection by qualified engineer
	Estimate subsoil profile from existing data and borings				No filed measurements

(continued)

Table 15.7   (continued)

III Average					Maintenance limited to emergency repairs	
	0.6	0.6	0.6	0.6	0.6	
IV Poor	No field investigation	No laboratory tests on samples obtained at the site	Approximate analyses using assumed parameters	No construction supervision by qualified engineer / No construction on control tests	Occasional inspection / No field measurements	
	0	0	0	0	0	

15.5.5   Categories, FoS and Probabilities of Failure of the Dams

Using the Categories from both the SLM method and ORE2_Tailings together with the results (Table 15.4) of the engineers' stability analyses for the various dams of the portfolio, Table 15.9 can be built, using the curves of Fig. 11.1. Values are cut at credibility threshold of one in a million or 10^{-6} (see Sect. 7.3).

As can be seen from Table 15.9, short boreholes (Dams 1A and 1B) deliver higher probabilities of failure to a dam that is overall well-designed, built, maintained and managed under constant FoS. Needless to point out that this corresponds to observations in various recent catastrophic failures. The same occurs (Dam 3) if liquefaction is suspected, again in compliance with recent observations.

Also, from Table 15.9 it is easy to see that the difference between "optimistic" and "pessimistic" category classifications delivers probabilities of failure which lie within one order of magnitude or less. That variation is compatible with extant uncertainty about the performance of the world-wide portfolio over the last hundred years, leading to the conclusion that the portfolio prioritization does not require systematic optimistic/pessimistic range evaluation but, more importantly, requires consistency in the evaluation for each single item.

For Dam 4, Sect. 11.1.2 reported a probability of failure between 6 and 13% based on the sole analysis of Eq. 11.4 and some simple considerations related to the empirical distribution of the FoS. The p_f evaluations in Table 15.9 yield higher values (19–38%) because they take into account all the uncertainties, lack of monitoring, repairs and maintenance surrounding this dam, and in particular the uncertainties on the geotechnical investigations that are not at all included in the FoS determination.

A careful reader will see from Table 15.9 that, generally speaking, the SLM method and ORE2_Tailings deliver similar values of the categories for the portfolio's

Table 15.8 SLM (Silva et al. 2008) results for Dam 4

Cat.	Design			Build	Operate	
	D1	D2	D3	CO	OM	
	Investigation	Testing	Analyses and documentation	Construction	Operations and monitoring	Category
I Best	Evaluate design and performance of nearby structures	Run lab tests on undisturbed specimen at field conditions	Determine FS using effective stress parameters based on measured data for site	Full time supervision by qualified engineer	Complete performance program including comparisons between predicted and measured	3 to 3.2
	Analyze historic aerial photos	Run strength tests along field effective total and effective stress paths	Consider field stress path in stability determination	Construction control tests by qualified engineers and technicians	No Malfunctions (cracks slides etc.)	
	Locate all non uniformities	Run index field tests (field vane, come penetrometer) to detect anomalies	Prepare flow nets for instrumented sections	No errors or omissions	Continuous maintenance	
	Determine site geologic history	Calibrate equipment and sensors prior to tests	Predict pore pressure and other relevant performance parameters for instrumented sections	Construction report clearly documents construction activities		
	Determine subsoil profiles using continuous sampling		Have design report clearly document parameters and analyses used for design			
	Obtain undisturbed samples for testing		No errors or omissions			
	Determine field pore pressure		Peer review			

(continued)

Table 15.8 (continued)

	0	0	0	0	0	
	Evaluate design and performances of nearby structures	Run standard tests on undisturbed specimen	Determine FS using effective stress parameters and pore pressure	Part time supervision by qualified engineer	Periodic inspection by qualified engineer	
	Exploration program tailored to project conditions by qualified engineer	Measure pore pressure in strength tests	Adjust for significant differences between field stress paths and stress path implied in analysis that could affect design	No errors or omissions	No uncorrected malfunctions	
		Evaluate differences between laboratory tests conditions and field conditions			Selected field measurements	
II Above Average					Routine maintenance	
	0.4–0.5	0	0.4–0.5	0	0	
	Evaluate performances of nearby structures	Index tests on samples from site	Rational analyses using parameters inferred from index tests	Informal construction supervision	Annual inspection by qualified engineer	
	Estimate subsoil profile from existing data and borings				No filed measurements	
III Average					Maintenance limited to emergency repairs	
	0.4–0.5	0	0.4–0.5	0.6	0	
	No field investigation	No laboratory tests on samples obtained at the site	Approximate analyses using assumed parameters	No construction supervision by qualified engineer	Occasional inspection	
				No construction control tests	No field measurements	
IV Poor	0.4–0.5	0.8	0.4–0.5	0.6	0.8	

Table 15.9 Annual probabilities of failure for the Dams in the portfolio evaluated using SLM and Ore2_Tailings approaches

Dam	FoS			CAT. SLM	Annual p_f		
	USA	ESA	Pseudo-static		USA/rapid draw down	ESA	Pseudo-static per event
1A	n/a	1.61–1.68	1.18–1.23 return 1/475	1.4	n/a	$\leq 10^{-6}$	$2.03 * 10^{-3}$ to $5.6 * 10^{-3}$
				1.6	n/a	$\leq 10^{-6}$ to $2 * 10^{-6}$	$4.83 * 10^{-3}$ to $1.16 * 10^{-2}$
				1.45	n/a	$\leq 10^{-6}$ to $1.2 * 10^{-6}$	$2.46 * 10^{-3}$ to $6.71 * 10^{-3}$
				2.22	n/a	$2.61 * 10^{-5}$ to $7.8 * 10^{-5}$	$2.99 * 10^{-2}$ to $6.55 * 10^{-2}$
1B	n/a	1.59–1.65	1.15–1.26 return 1/475 and 1/1500	1.4	n/a	$\leq 10^{-6}$ to $1.36 * 10^{-6}$	$1.1 * 10^{-3}$ to $1.03 * 10^{-2}$
				1.6	n/a	$1.26 * 10^{-6}$ to $4.03 * 10^{-6}$	$2.45 * 10^{-3}$ to $2.07 * 10^{-2}$
				1.45	n/a	$\leq 10^{-6}$ to $1.79 * 10^{-6}$	$1.35 * 10^{-3}$ to $1.23 * 10^{-2}$
				2.22	n/a	$4.71 * 10^{-5}$ to $1.74 * 10^{-4}$	$1.87 * 10^{-2}$ to $1.05 * 10^{-1}$
2	Rapid draw-down 1.63–1.76	1.59–1.68	1.28–1.33 return 1/1,000	2.2	$6.53 * 10^{-6}$ to $5.12 * 10^{-5}$	$2.32 * 10^{-5}$ to $9.65 * 10^{-5}$	$5.93 * 10^{-3}$ to $1.31 * 10^{-2}$
				2.4	$2.44 * 10^{-5}$ to $2.64 * 10^{-4}$	$7.49 * 10^{-5}$ to $2.64 * 10^{-4}$	$1.01 * 10^{-2}$ to $2.32 * 10^{-2}$
				2.17	$5.36 * 10^{-6}$ to $4.36 * 10^{-5}$	$1.95 * 10^{-5}$ to $8.3 * 10^{-5}$	$5.48 * 10^{-3}$ to $1.23 * 10^{-2}$
				2.36	$1.88 * 10^{-5}$ to $1.22 * 10^{-4}$	$5.92 * 10^{-5}$ to $2.16 * 10^{-4}$	$9.07 * 10^{-3}$ to $1.86 * 10^{-2}$
3	1.6	1.29–1.35	1.2 return 1/2,475	1.6	$3.32 * 10^{-6}$	$4.64 * 10^{-4}$ to $1.37 * 10^{-3}$	$7.84 * 10^{-3}$

(continued)

Table 15.9 (continued)

		FoS			
		1.7	$6.74 * 10^{-6}$	$6.60 * 10^{-4}$ to $2.06 * 10^{-3}$	$1.14 * 10^{-2}$
		1.83	$1.17 * 10^{-5}$	$1.16 * 10^{-3}$ to $3.52 * 10^{-3}$	$1.84 * 10^{-2}$
		2.37	$1.97 * 10^{-4}$	$7.00 * 10^{-3}$ to $1.65 * 10^{-2}$	$5.96 * 10^{-2}$
4	Refer to Tables 11.1 and 11.2 in Sect. 11.1.2: FoS average 1.17 used in p_f evaluations	3	n/a	$1.94 * 10^{-1}$	n/a
		3.2	n/a	$2.30 * 10^{-1}$	n/a
		3.3	n/a	$2.50 * 10^{-1}$	n/a
		3.8	n/a	$3.80 * 10^{-1}$	n/a

dams. However, the SLM method is a much more deterministic than ORE2_Tailings in the sense that the range of probability of failure is narrower. Also, for a few cases the SLM method gives completely different values, as it lacks critical data input specific to tailings dams.

Finally, it is easy to verify that the annual probability of pseudo-static failure—i.e., the pseudo-static p_f divided by the return—is in the particular case of this portfolio always either lower than or of the same order of magnitude as the p_f ESA (see Sect. 11.2.3 for some caveats).

Thus, in Fig. 15.6 we have only included the data of ESA with the probability of failure as a function of a FoS. Figure 15.6 shows a comparison between the SLM and ORE2_Tailings results for the probability of failure as a function of the FoS determined by the engineers (Table 15.6) for each dam of the portfolio.

As already stated earlier regarding Dams 1A and 1B, it is worth noting that the SLM method does not contain any provision for water balance or freeboard, as it is a general slope methodology, while ORE2_Tailings includes those factors and many more, dam-specific ones. ORE2_Tailings compute those probabilities mainly because of the short, relatively widely-spaced boreholes performed during the investigations. The same is true for for Dam 1B (see Sect. 15.5.1 for more details).

In the case of Dam 2, where there are few tailings dam-specific features, the correspondence of SLM with ORE2_Tailings is very good.

For Dam 3: the SLM method does not consider liquefaction, the divergence with ORE2_Tailings here is again significant (see Sect. 15.5.3 for more details).

For Dam 4: ORE2_Tailings penalizes the probability of failure as it is characterized by the paucity of data and ignorance on many aspects of its history.

The SLM method does not look at specific tailings dams deficiencies.

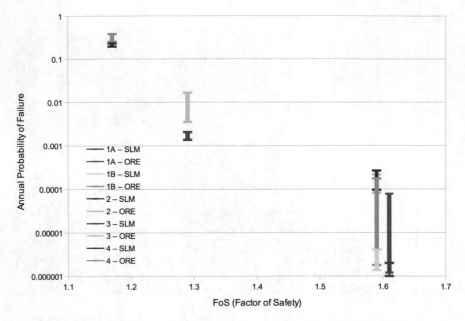

Fig. 15.6   SLM versus ORE2_Tailings FoS versus Annual probability of failure for ESA

Similar conclusion can be derived from USA rapid drawdown and pseudostatic conditions.

Now, let's go back and look at the areas where further analysis is required (pipes, traffic, etc.).

Internal interdependencies

- We stated earlier (Sect. 15.1.3) there are no penstocks systems in the portfolio's dams, with the exception of Dam 4 and that is already included in the deplorable category of that dam.
- There are tailings pipelines at the crest of Dams 1A, 1B (insensitive to pipe breach), and 3.
- Dam 3 could be significantly eroded if the crest pipeline were to breach and remain unobserved for more than 12 h (Sect. 15.5.3).
- Dam 4 is in the process of deactivation, thus no active spigotting.
- There are vehicular and subcontractor's traffic at the crest of all dams.

For the sake of the example, let's look at Dam 3 pipeline breach and traffic case.

As shown in Fig. 8.8, a breach in a crest pipeline can lead, by internal inter-dependencies, to the collapse of the dam. This requires a chain of events, for example, as follows:

- The pipeline breaches due to: (a) a vehicle collision, (b) a joint failure, (c) corrosion of the pipe. As each one of these three failure modes can be considered

Table 15.10 External interdependencies

Dam 1	Dam 2	Dam 3	Dam 4	Alters the probability (and consequences) of a failure of
0	n/a	n/a	n/a	Dam 1
n/a	0	n/a		Dam 2
n/a	Could trigger liquefaction or instability of Dam 3	0	n/a	Dam 3
n/a	n/a	n/a	0	Dam 4

independent from the others, Sect. 4.2.3 and in particular Eq. 4.2 yield the probability of failure as

$$p_{f(a,b,c)} = 1 - \left[(1 - p_a) * (1 - p_b) * (1 - p_c) \right].$$

- If the pipeline breach goes undetected (for this dam for more than 12 h) and the crest design is such that a spill towards the downstream face of the dam is not hindered, for example by a berm, then the spill may start scouring the face and the tailings will start filling the next berm surface.
- If more time goes by and still no detection occurs (for example by a patroller), then scouring may start …
- which may then evolve into sloughing and erosional instabilities …
- which may then lead to a dam global instability.

Event Trees Analyses (ETAs) or Failure Tree Analyses (FTAs) may be prepared to model this development. As stated in Sect. 8.3.1 these analyses require the evaluation of elemental probabilities which can be based on Appendix A.

It is therefore evident that this type of approach becomes convergent, insofar as risks from multiple hazard types (natural, man-made, technological, etc.) can be evaluated simultaneously and yet be extracted by queries to study their specific impacts on the risks of the portfolio, provided the hazard and risk register is designed with this purpose in mind.

External interdependencies

Dams 1 and 4 have no significant internal or external interdependencies.

Dam 3 has internal interdependencies (progressive failure). The probability of failure of Dam 3 would be altered by the failure of Dam 2 and the consequences of a compound failure would be worse than the consequences of the failure of either Dam 2 or Dam 3 alone.

Table 15.10 explains the interdependencies between different dams. For this case study, where only one interdependency exists, it might seem overly complicated to prepare such a table, however it is the only way to ensure completeness and concision.

The internal interdependencies can be noted in Table 15.10. As an example, the internal interdependencies of Dam 3 are noted by inserting -1- in the cell on the diagonal for Dam3-Dam3.

As can be inferred from the discussion above, no dam failure in this portfolio could be called unforeseeable, with an exception made for some extreme evaluations of Dams 1A and 1B. As stated in Sect. 7.3, we shall not call unforeseeable what is indeed foreseeable.

The dams of the portfolio are all in the Cassandra category, except for Dam 3 which we would categorize as Pythias (Sect. 7.4). Perhaps Dams 1A and 1B could be considered Sword of Damocles by some readers, although the shallow depth of their geotechnical investigations tells us the contrary. Finally Dam 4, a catastrophe waiting to happen, could also be classed as a Medusa.

15.6   Consequences

Going back in time, in Oboni et al. (2013) we stated that: "Especially for very large projects, risk assessments generally consider too simplistic consequences and ignore "indirect/life-changing" effects on population and other social aspects" that inevitably will hinder SLO/CSR. Using conceptually sound risk assessment methodologies makes it possible to consider the significant uncertainties that surrounds the driving parameters and related consequences. Among these (see Sect. 12.1):

- human H&S (health and safety);
- fish, fauna and top-soil/vegetation consequences;
- long term economic and development consequences, and social impacts.

Similarly, other recommended consequence models should include (see Sects. 7.4 and 7.6):

- extent of damage as expressed in casualties, business interruption, etc. possibly merged into one metric, with the inevitable uncertainties;
- geographic dispersion of damage;
- duration of the damage;
- reversibility of the damage (or perpetual losses?);
- latency between an accident and the occurrence of its damages;
- social impacts (inequity/injustice, psychological stress and discomfort, conflicts and mobilization, spillover effects) (see Sect. 5.5).

The Mackenzie Valley Review Board (MVRB 2013: Appendix D, see Sect. 5.2) defined what a modern risk assessment should include, based on the results of public hearings. Thus it becomes clear that partial components of the consequences such as:

- biological impacts and land use
- regulatory impacts and censure
- public concern and image
- health and safety (H&S)
- direct and indirect costs

have to be included.

Since risk data are often highly uncertain, some people are likely to object that data and engineering judgments are too divergent to trust. In addressing this issue, several guidelines can be helpful (for more details, see Hance et al. 1987).

Based on the system map (Fig. 15.5) the following consequences are evaluated:

Dam 1A: Extensive damages to the town and town infrastructure, and tailings runout to the river stream with river transportation mechanism.
Dam 1B: Full loss of the town and town infrastructure, and tailings runout to the river stream with river transportation mechanism.
Dam 2: water flooding into the valley with inundation of the low areas of the town and industries, and loss of service of transport infrastructures.
Dam 3: tailings runout blockage in the valley, then flooding in the low areas of the town's industries and loss of service of transport infrastructures.
Dam 4: (Sect. 11.2.1) damages in the mine perimeter, road failure, personnel safety, mining infrastructure.
Interdependent failure of Dams 2 and 3: The combination of water and tailings makes the tailings more fluid; important scouring occurs; wider areas are impacted because of the combined volume and higher energy.

Based on the above and on the definition of consequences we delivered in Sect. 7.6, and Chap. 12, the dimensions of the failure selected for this discussion are as follows:

PL   physical losses
HS   health and safety damages
BI   business interruption
ED   environmental damages
RD   reputational damages, legal fines, public outcry (community outrage) damages
CR   crisis potential

where:
PL = value of significantly damaged assets including third-party assets. The "cost of the dam" will be excluded from this discussion. To include it we should understand how the structure is insured and the discussion would expand beyond the scope of this book.

HS = evaluation of the global potential health and safety cost using the WTP (see Sect. 13.1.2) assumed in this discussion at 2.5 M$.

BI = for small failures the business interruption is evaluated in weeks or month for the clean-up to take place and appropriate infrastructure replaced. However, for catastrophic failure or a failure that will cause third-party casualties—i.e., failures that will cause a crisis potential—the BI will be evaluated in years (including legal shutdowns etc.) until the operation can reopen. Mount Polley (see Sect. 2.1) took two years to reopen and Samarco (see Sect. 2.2) is still closed four years after the catastrophe. The resilience toward BI damage depends on the company that owns/operates the failed dam. A company with a large portfolio of mines will be less sensitive to the impact, capable of buffering against BI by increasing production at other sites, hence the overall damage may be greatly mitigated. A company with a small portfolio of mines will receive the full impact from the BI of one dam. Thus, discussing the

Table 15.11 Quantified consequences for each dam

Dam	PL (M$)	HS (casualties)	BI (years)	ED (M$)	RD
1A	35	40	3	100	5
1B	90	100	5	200	10
2	10	10	0.5	10	2
3	15	20	2	60	5
4	8	10	2	120	3
2 on 3	40	50	3	160	10

BI damage requires an enterprise risk management approach, where interdependencies are looked at a higher level, still approachable with the type of rational analysis described in this book. However, due to obvious limitations of space, we will exclude BI from this discussion.

ED = include various sub-dimensions: geographic dispersion of damage, duration of the damage, reversibility of the damage (see Sect. 7.4). In some studies, the environmental damages are evaluated in classes as a simplification, but the preferred practice is to analyze each sub-dimension.

RD = measures the reputational damages and public outcry, including legal fines, with a "consequences multiplier." A range between 1 and 10 seems reasonable based on our experience. Numerous recent examples ranging from the Mont Blanc tunnel to Fukushima to the Lac Megantic rail disaster and others have shown that the approach of "fact-driven" approach to the evaluations of consequences will lead its user to an unsustainable stance (https://www.riskope.com/wp-content/uploads/2013/10/Quantifying-Social-Perception-of-an-Industrial-Accident-Risk.pdf).

CR = impact on SLO and CSR, considered here to impact BI (boycotts, blockades, etc.) as a multiplier of the BI duration.

The additive consequence dimensions can be evaluated as shown in Table 15.11.

As the BI damage will not enter in the discussion (see Table 15.11 and explanations above), the multiplication of BI by CR will also be excluded from further analyses in this discussion.

15.7   Portfolio Risk Assessment and Benchmarking

Now that probabilities and consequences (ranges) are evaluated it is possible to determine risks for each dam using the definition of risk we gave in Sect. 1.2.

As discussed in Sect. 12.1 the consequence function is additive, i.e., a sum of the various consequence dimensions. For HS (casualties) we have selected a WTP = 2.5 M$ value as an average (see Sect. 13.1.2). Two values of C are calculated in Eqs. 15.1 and 15.2, C_{min} and C_{max}, respectively without and with the influence of RD:

$$C_{min} = (PL + (HS * 2.5) + ED) \qquad (15.1)$$

$$C_{max} = (PL + (HS * 2.5) + ED) * RD \qquad (15.2)$$

Table 15.12 displays, for each dam and for the scenario of Dam 2 interdependent on Dam 3, the risk evaluation for various cases of p_f (ESA) and C_{min} and C_{max}. The italic rows correspond to the lower estimates of the p_f (ESA). In other words, the italic rows correspond to the optimistic category classification of Table 15.9, whereas the rows in roman correspond to the pessimistic category classification. Thus, for each dam four risk estimates are generated.

If the failure criteria (see Sect. 7.7) has been carefully selected, bench-marking becomes a powerful tool of comparison. Whereas risk comparison amounts to showing different activities of comparable risks, benchmarking corresponds to rating a given activity's risk related to other similar activities. Benchmarking of a tailings dam portfolio's probability of failure can assume the aspect shown in Fig. 15.7 (Oboni and MDA 2018).

The extremes of the bars represent the minimum and maximum estimates of the probability of failure for each dam. Hence the length of the bars measures the uncertainty on the evaluation of the probability at the time of the assessment and with the data available at that time. The vertical axis shows four benchmark values (green), namely:

- Mount Polley and Samarco annual p_f evaluated with data available to us BEFORE the respective failures;

Table 15.12 Risk evaluation for various cases of p_f (ESA) and C_{min}, C_{max}

Dam	p_f (ESA)		Consequences		Risk	
	min	max	min	max	min	max
1A	$1.00 * 10^{-6}$	$1.20 * 10^{-6}$	235	1175	$2.35 * 10^{-4}$	$1.41 * 10^{-3}$
1A	$2.61 * 10^{-5}$	$7.80 * 10^{-5}$	235	1175	$6.13 * 10^{-3}$	$9.17 * 10^{-2}$
1B	$1.00 * 10^{-6}$	$1.79 * 10^{-6}$	540	5400	$5.40 * 10^{-4}$	$9.67 * 10^{-3}$
1B	$4.71 * 10^{-5}$	$1.74 * 10^{-4}$	540	5400	$2.54 * 10^{-2}$	$9.40 * 10^{-1}$
2	$1.95 * 10^{-5}$	$8.30 * 10^{-5}$	45	90	$8.78 * 10^{-4}$	$7.47 * 10^{-3}$
2	$5.92 * 10^{-5}$	$2.16 * 10^{-4}$	45	90	$2.66 * 10^{-3}$	$1.94 * 10^{-2}$
3	$1.16 * 10^{-3}$	$3.52 * 10^{-3}$	125	625	$1.45 * 10^{-1}$	$2.20 * 10^{0}$
3	$7.00 * 10^{-3}$	$1.65 * 10^{-2}$	125	625	$8.75 * 10^{-1}$	$1.03 * 10^{1}$
4	$2.50 * 10^{-1}$	$2.50 * 10^{-1}$	153	459	$3.83 * 10^{1}$	$1.15 * 10^{2}$
4	$3.80 * 10^{-1}$	$3.80 * 10^{-1}$	153	459	$5.81 * 10^{1}$	$1.74 * 10^{2}$
Dam 2 on 3	$7.80 * 10^{-6}$	$3.32 * 10^{-5}$	325	3250	$2.54 * 10^{-3}$	$1.08 * 10^{-1}$
Dam 2 on 3	$2.37 * 10^{-5}$	$8.64 * 10^{-5}$	325	3250	$7.70 * 10^{-3}$	$2.81 * 10^{-1}$

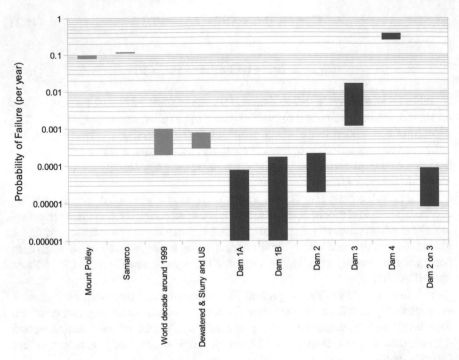

Fig. 15.7 Annual probability of failure versus dams of the portfolio and benchmarking cases (green). The annual probability of failure for each dam is displayed as an orange-red bar due to the uncertainties. The longer the bar, the larger the uncertainties

- the min-max values of the world-wide p_f portfolio based on historic records (Oboni and Oboni 2013), represented by the green bar "World-wide decade around 1999";
- the values of p_f ("Dewatered and slurry and US", green bar) obtained in Taguchi (2014), which attempted a theoretical estimate of the annual p_f of standard and dewatered tailings.

The portfolio benchmarking in terms of the probability of failure shows that in the considered case there are dams (1A, 1B, 2, 2 on 3) below the historic benchmark. Dams 3 is above the historic benchmarks, although still significantly lower than the Mount Polley or Samarco estimates made with available pre-failure data. Dam 4 is an imminent catastrophe. Additional studies and information would make it possible to reduce the uncertainties (length of the orange-red bars).

Mitigation would push the bar down, an aspect we will discuss later in this chapter. Long term lack of maintenance and the effects of climate change will tend to push the bars up, as has already occurred with Dam 4.

15.8   Risk Tolerance and Acceptability Thresholds

In Fig. 15.8 an orange line represents the corporate tolerance curve (see Sect. 13.2.1) selected by the owner of the portfolio. For this particular corporation a 100 M$ loss event with a frequency of one out of ten years is intolerable. Also intolerable is an event generating a loss of 4500 M$ (4.5 B$) with a probability at the limit of credibility (10^{-6}).

In Fig. 15.8 the yellow squares depict the C_{min} scenarios (i.e., without the RD) and the optimistic probability of failure (p_{fmin} and C_{min}) for each dam. Only Dam 4 is intolerable under these conditions. However, as history has demonstrated in various occurrences, neglecting the reputational damages is a flawed and unrealistic approach.

Figure 15.9 only displays the blue squares, representing for each dam the c_{max} and the pessimistic probabilities of failure (p_{fMax} and C_{max}).

As discussed above, out of prudence, from this point on we will focus the discussion on the results displayed in Fig. 15.9.

We immediately see that Dam 1A, Dam 2 and Dam 3 are below the selected corporate tolerance. Also, Dam 4, Dam 1B, and Dam 2 interdependent on Dam 3 are all above tolerance.

Figure 15.10 displays the total risk for each structure as the sum of the tolerable part—i.e., the portion below tolerance (blue)—and the intolerable portion (orange)—i.e., the portion of the risk above the tolerance.

Figure 15.10 shows that prioritizing a portfolio mitigation plan based on total risk (blue + orange)—i.e., without considering the corporate tolerance to risk—would be far than optimal and lead to squandering of mitigation capital because:

- Dam 3, tolerable, would be mitigated before Dam 1B, which has an intolerable portion of risk, and

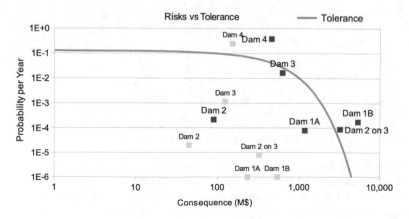

Fig. 15.8 In blue p_{fMax} and C_{max} for each dam, and in yellow p_{fmin} and C_{min} for each dam. Corporate tolerance is displayed in yellow

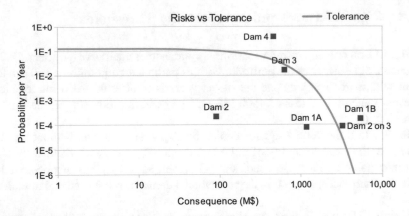

Fig. 15.9 The blue squares represent for each dam the C_{max} and the pessimistic probabilities of failure (p_{fMax} and C_{max}). Corporate tolerance is displayed in orange

- Dam 3 (tolerable) would be mitigated before the interdependence of Dam 2 on Dam 3, which has an intolerable portion of risk, despite its probability being two orders of magnitude lower.

From the above we note that risks, and in particular risk prioritizations, cannot be defined by gut-feeling and intuition. The reason is simple: our human brain already has enough trouble understanding the simultaneous effect of two parameters (i.e., probability of failure and cost of consequences) and the introduction of a third vital one (the tolerance) certainly does not help the brain to make optimal decisions. Thus, it is paramount to use rational approaches like the one discussed in this book to make decisions involving risks, even for simple portfolios of four dams such as the one discussed here, where decisions seem obvious!

> The phenomenon, highlighted in (Kahneman and Tversky 1979), is the "availability heuristic". It is one of the very well-known cognitive biases that plague us humans when we are confronted with decisions under uncertainty. The root cause of these biases are prejudices and misconceptions. Indeed, we often assess the probability or the magnitude, of an event by asking ourselves if there are "cognitively available" examples. Those are readily available through memory as Daniel Kahneman (Nobel Prize in Economics) and Amos Tversky demonstrated in a series of papers published between 1971 and 1984.

Given the legal aspects linked to tailings dams failures, while staying away from any legal discussion, we note that Figs. 15.9 and 15.10 become a powerful tool for the so-called test of negligence and decision making. Below is a summary of the test of negligence as practiced in various countries around the world.

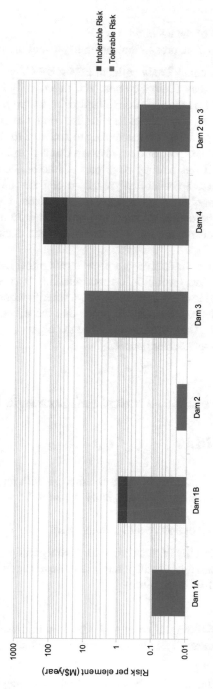

Fig. 15.10 Total risk for each dam is equal to the sum of tolerable (blue) and intolerable (orange) risks

Consider:

p_f = Probability failure

C = Cost of consequences of the event with p_f

M = Mitigative investment, i.e., precautions taken to prevent the risk (p_f * C)

The owner of the risk is considered negligent if M < P * C.

In other words, a judge may deem a company negligent only if mitigative moneys spent (per annum) are less than the annualized risks. Clearly, transparency and rationality constitute a strong a priori defense in case something should go wrong. Actually, going back to Fig. 15.10, it is interesting to note that the test of negligence would induce the owner of the portfolio to invest in Dams 3 and 4 first and then in Dam 1B, whereas Dam 1B has a fairly large part of intolerable risks. Examining Fig. 15.9 it is evident that the probability of failure of Dam 1B is two orders of magnitude lower than the p_f of Dam 3. Despite this, the existence of intolerable risks in Dam 1B would incite the owner to invest in Dam 1B before mitigating Dam 3. Later we will look at the societal aspects of these failures and decide if this decision meets the societal test.

Adding this rational and quantitative approach to risk prioritization brings value to the discussion and adds "foreseeability and controllability" to management's decisions. As a matter of fact, it is often impossible to act simultaneously on all required mitigations. Thus, it might be necessary to prove that the efficacy and efficiency were considered. Of course, the company would be the object of intense media and regulatory scrutiny should an accident occur.

15.9 Defining Operational, Tactical and Strategic Risks

15.9.1 Operational Risks

Dam 4 can be mitigated—i.e., brought under tolerance—with minimal effort at the operational level. A reduction of the probability by only 1.5 order of magnitude would make it tolerable.

In the case of Dam 4 two options could be attempted:

- reducing uncertainties and thus decreasing the appalling category of the structure, i.e., intensive monitoring, repairs, and possibly historic research and investigations (with care taken not to trigger a failure)

and/or

- increasing the FoS by building, if space and other conditions allow, a berm. The final choice should be carefully examined using probabilistic cost estimates (see Sect. 15.10).

These alternatives should be carefully weighed against the negligence test, which would require a 100 M$/year investment to pass. Clearly in the case of Dam 4,

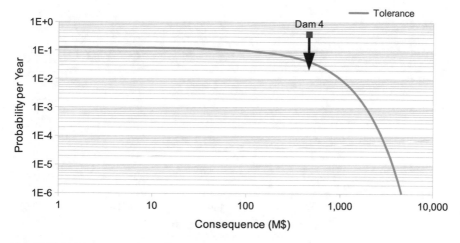

Fig. 15.11 Mitigation on dam 4 effects on risks

immediate action would be the best course. Any action cited above would lead to the black arrow in Fig. 15.11.

We will also note that Dam 3, with the highest tolerable risk of the portfolio, could be mitigated under operational conditions, and that this mitigation would serve operational goals as well as the tactical ones described in the next section.

15.9.2 Tactical Risks

The interdependencies of Dam 2 on Dam 3 can be lowered below tolerance by effecting some kind of mitigation on Dam 3 that would change the likelihood of its failure in case Dam 2 failed. An example of such a mitigation is an armored toe, preventing erosion of Dam 3, after careful consideration of the liquefaction potential.

Such a mitigation would be perfectly in line with the negligence test result, leading to a tactical investment of a few million dollars per year. Here again the portfolio risk assessment makes it possible to allocate resources and avoid paralysis caused by analysis and feeling overwhelmed (Fig. 15.12).

15.9.3 Strategic Risks

Dam 1B is considered intolerable independently of possible implementable mitigations (the black arrow in Fig. 15.13 remains above tolerance all the way down to the human credibility threshold). Dam 1B therefore represents a strategic risk; i.e., only an alteration of the system would make it tolerable (red arrow). Possible solutions

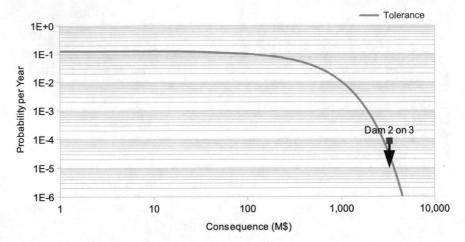

Fig. 15.12 The effects of mitigation on the risks of the interdependence of Dam 2 on Dam 3

might include moving the population, enhancing the protection to the environmental damages, lowering the dam, etc.

Here the test of negligence would indicate a necessary investment of $1 M/year, because the factual risk is small. However, the intolerable part of the risks, and the fact they are of a strategic nature, would induce the owner to mitigate more.

Thanks to the quantitative approach strategic shifts can be evaluated. We will discuss in the next Sect. 15.9.4 the potential impacts on human life and see how these would alter these initial considerations.

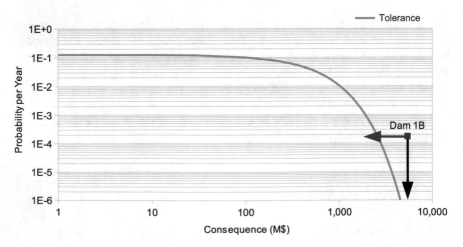

Fig. 15.13 Effects on risks of mitigation on Dam 1B

15.9.4 Impacts on Humans

Figure 15.14 displays the dam portfolio risks in terms of potential loss of lives in relation to Whitman's (1984) societal tolerance threshold and other industrial risks collected by Whitman. Both Dam 4 and Dam 3 have risks that are societally unacceptable. Going back to the considerations made earlier, Dam 3 is societally unacceptable with the result of placing it ahead of dam 1B.

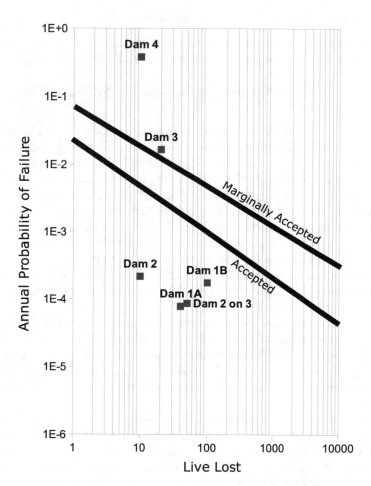

Fig. 15.14 Probability-live lost graph compared to Whitman (1984) thresholds

15.10   Risk Informed Decision Making and Roadmap

What has been discussed up to this point offer firm support for risk-informed decision making. RIDM is necessary to prioritize dams in order of urgency and decide which mitigations to implement. If we go back to Sect. 1.1 where RBDM and RIDM were defined, RBDM would be used to prioritize actions and select mitigations, but it would not be sufficient to evaluate the effectiveness of the mitigative investment in order to optimize it.

When it comes to mitigation, actions need to be:

- efficient, that is, work in a quick and organised way (covered by RBDM);
- efficacious, i.e., able to produce the intended results (covered by RBDM) and finally,
- effective, meaning an action giving the desired results (within clear financial goals, covered by RIDM).

The idea is to develop the implementation of effective risk reduction strategies using RIDM. Especially when it comes to comparing capital expenditures and long-term monitoring, risk-informed decision making requires more sophisticated methods than the classic financial tools like Net Present Value (NPV) or the mere "counting of risks" carried out using traditional matrix-based risk assessments, due to the inherent and insurmountable limitations (https://www.riskope.com/2009/08/19/stop-procrastinating-npv-is-dead-use-risk-as-a-key-decision-parameter/) those approaches have (Robichek and Myers 1966). Our experience has shown that limitations can be overcome with economical (https://www.sapling.com/6708011/calculate-riskadjusted-npv) and sustainable methodologies which can deliver a significant competitive edge to their users (Espinoza 2015).

Leaving aside the details of the financial evaluations, the first step of RIDM is to gain an understanding of which risks are operational, tactical and strategic, and societally acceptable (see Sect. 15.9) and which levels of mitigation are dictated by the negligence test (see Sect. 15.8). This classification is paramount in this phase; it is an important result of a modern risk assessment, and not the result of an arbitrary selection.

For the case study discussed in this chapter the synthesis of these evaluations is displayed in Table 15.13, where we focus on the dams with intolerable or almost intolerable risks (Fig. 15.9) and societally unacceptable dams (Fig. 15.14). Interestingly, Dam 2 by itself is fully tolerable, but when seen in the scenario of its being interdependent on Dam 3, it is not.

From Table 15.13 we can distil the following roadmap, which would still require a careful risk-adjusted cost analysis and evaluations of residual risks.

- First priority, Dam 4. Every effort must be made to reduce corporate, legal and societal liability and exposure.
- Second priority, Dam 3. Although this dam is corporately tolerable, it sits right at the threshold. Its negligence test has a significant yearly value and it is societally

Table 15.13 Synthesis of the results of the portfolio analysis developed in this chapter

Evaluation criteria	Dam			
	1B	2 on 3	3	4
Corporately intolerable (qualitative Fig. 15.9)	Yes	Barely	Almost	Widely
Corporately intolerable (quantitative Fig. 15.10) (M$/year)	0.5	<0.1	Largest tolerable	80
Negligence test min. mitigation (M$/year)	1	0.2	10	100
Societally unacceptable (Fig. 15.14)			Yes	Widely
Mitigations				
Operational (Fig. 15.11)			Reinforce toe to be evaluated	Buttress and/or reduce uncertainties or other to evaluate
Tactical (Fig. 15.12)		Reinforce Dam 2 or toe of Dam 3, or both to evaluate		
Strategic (Fig. 15.13)	Move population, lower the dam to evaluate			

unacceptable. Moreover, the operational mitigation would "kill two birds with one stone" as it would also mitigate the scenario of Dam 2 interdependent on Dam 3.

- Third priority, Dam 2 interdependent on Dam 3. This interdependent scenario is barely corporately intolerable and has a low negligence test. Possible mitigation must be seen in conjunction with interdependent Dam 3.
- Fourth priority, Dam 1B. Although this dam represents a strategic risk the low probability of failure, its societal acceptability, low negligence test value and low intolerable risk suggest the lowest priority in the portfolio.

After residual risk evaluations, i.e., the evaluation of the entire portfolio including the selected mitigative measures it would be necessary to repeat the roadmap exercise.

15.11 Risk Communications

In Chap. 5 we discussed "what people want". With the results of the risk assessment at hand we can now test and return to the discussion of risk communication.

Enhanced transparency, sensible information, and better risk communication will certainly help to foster robust corporate social responsibility (CSR) and social license to operate (SLO) in addition to fulfilling societal requirements (Chap. 5) and fostering healthy NI43-101 compliance. Breaking the siloes between risk assessment, SLO and CSR becomes paramount to maintaining sustainable operations and projects.

In the following pages we tackle specific aspects of risk communication, risk acceptance/tolerance (see Chap. 13) and risk comparisons. We will see that fostering rational quantitative, updatable and convergent risk assessments is of paramount importance to support not only technical aspects of risk management, but also to boost any CSR, SLO and NI43-101 activity.

Indeed CSR approaches have inherent limitations when they:

- treat CSR as public relations;
- focus on the system's elements that generate the largest risks rather than on those (people, environments) with the greatest need;
- excessively simplify complex processes;
- focus on maintaining the status quo.

These limitations hinder proper consideration for those living nearby a mining operation, even when CSR activities are considered a corporate priority.

15.11.1 Enhancing Transparency Through Risk Communication

Experience shows that successful risk communication should always focus on planned control measures and precautionary actions rather than on risks. As in many other fields, actions speak louder than words and are what matters to the public. Indeed, people demand to know what preventative actions are foreseen even if it is claimed that the likelihood of an accident and associated risk is very low. Experience has unfortunately shown that events with a very low likelihood of occurrence actually do occur (Fukushima and other nuclear accidents, incidents in the hazardous materials industry, etc.). We are not mentioning tailings dams breaks here, as at a rate of one to four failures per year (Oboni and Oboni 2013), no one should actually talk about a very low likelihood for this type of accident in the world-portfolio.

The considerations above may explain why the public:

- puts more emphasis on what is done to reduce mishaps, rather than on the details of the mishaps and their potential effects;
- cares about the skills and awareness of the operation manager more than about the risk itself;

- perceives mismanagement, incompetence, and lack of conscientiousness as the central issues in accidents (Bouder et al. 2007).

Moreover, academic and popular literature suggest agreement that the public's distrust of industry has developed over the past half-century as a result of repeated failures to provide adequate and/or accurate risk information to the public and, in some cases, blatant or perceived conflict of interest (Oboni et al. 2013).

At the other end of the spectrum of inaccurate information we find sensationalism, especially if unsupported by facts, and the result of biased interpretation of incomplete databases or "gut feeling" (see Sect. 15.8). This is obviously not a recommendable path when communicating risks. Considering that today modern dams, especially the largest ones, are better investigated, designed, built and managed/monitored that any older structure, based on our experience, we see their probability of failure decrease towards a value close to those of hydraulic structures of the same caliber. The portfolio studied in this chapter is an eloquent demonstration of this statement.

Just as a reminder, many catastrophic failures in the last 45 years occurred for medium-height dams, storing base and precious metals tailings. Samarco was an exception with its 100 m height, but its overall quality was poor, as highlighted by forensic studies, with design changes and an extremely high average raise per annum. The recently failed Corrego dam (near Bernardinho, Brazil) had the longest life to failure of the group (43 years), Samarco and Baia Mare the shortest. In all these failed structures, investigations, design, construction, operations and management lead them far astray from what would be considered excellent quality in a large dam today.

This leads to us recommend avoiding any sensationalist statements about risks, as the interplay of probability and consequences necessarily means neither that risks of a large dam are always bigger nor that smaller dams have more tolerable failures. Additionally, and to confirm the above, we unambiguously showed in (Caldwell et al. 2015) that the graph relating volume released to lives lost does not show any correlation. The same would be true of the dam height or the run-out distance or outflow volume with the same conclusions!

In summary, explaining what is planned, is now being done, and is foreseen to mitigate risks is at least as important as explaining how small those risks may be. The public may be much more receptive if mitigative plans come first and the statement "trust us, risks are small" is avoided, as we pointed out in Sect. 1.1. In conclusion:

- Simplistic statements equating bigger dams with bigger risks are to be left aside as they can easily be disproven and do not help any constructive dialogue.
- As structures become larger, their probability of failure should go down towards the values commonly accepted for major hydro-dams. That reduction necessitates action on all aspects of the dam investigation, design, construction, monitoring, inspections and management, as all contribute to the chance of failure. Increasing the FoS does not clearly reflect any of the above necessary improvements.
- Population increase, land use shifts, and environmental constraints increase the consequences of failures and therefore tend to increase risks. Again, the probability

of failure will have to go down, especially considering that tailings dams have a long life span during service and post-service, and, again, the FoS cannot help measure the changes.

15.11.2 Risk Communication Using Comparisons

Most tables of risk comparisons in the literature contain a mix of risks characterized by different levels of uncertainty. In addition, most risk comparison tables offer only single-digit risk estimates, with no range or error term. For risks such as driving, where fatalities can be counted, the number is likely to be reliable, at least in some countries. However, even if the risk comparison data are carefully and accurately reported, they can be misleading. For example, the risk statistics for driving includes many different driving situations. Yet speeding home after a long party in the early morning is two orders of magnitude more dangerous than driving slowly to work during the day. This type of reasoning may explain why in our experience people have trouble relating such "general" risks to the specific risks posed by a project of, say, a tailings dam.

As a result, risks such as those generated by major tailings dam failures, occurring at a rate of 10^{-3}–10^{-4} (Oboni and Oboni 2013; see Chap. 4) estimated on the basis of incomplete statistics and flawed by under-reporting and other biases, or necessarily simplified models and evaluation approaches, require great care in presentation and comparison. As already pointed out, consistent relative risk and benchmarking are of paramount importance, as are complete analyses of consequences and well-developed dam break studies.

Useful risk comparison should be accurate and pertinent. For example, comparing the risks of a chemical processing plant to voluntary actions such as smoking or driving without a seat-belt is neither accurate nor pertinent.

15.11.3 Anticipating Objections

A risk assessment (RA) should be performed by an independent entity (Brehaut 2017; Roche et al. 2017). Nevertheless, the proponent of a project and the public might have objections to the RA. In this section we discuss how to help remove objections in order to bring credibility and ultimately SLO and CSR by applying a mix of technical and soft concepts and skills.

Explicitly considering and explaining data uncertainties helps to explain how risk estimates are obtained and appease the objections summarized above. Enhancing the transparency of risk assessments is of course extremely important in this line of thought. This goal cannot be reached if, for example, arbitrary limits are selected for boiler-plate risk matrices where the coloring scheme does not bear any relation to public (or corporate) risk tolerance (Chapman and Ward 2011; Cox et al. 2005;

Hubbard 2009) and risk assessment is reduced to a binning exercise of "magic single numbers" like, say, likelihood = 3 * impact = 5, thus risk = 15! (see Sects. 1.2 and 4.2.1).

Avoiding misleading statements, openly exposing uncertainties, and abolishing arbitrary selections are all actions leading to demystifying the risk assessment process. They are there for the benefit of all parties. The public (and management) will indeed be in a better position to understand why risk estimates are fraught with inevitable uncertainties, despite the fact that they are based on the best available data and unbiased interpretation. That is, of course, provided conflict of interest and biases are kept at bay during the process.

Acknowledging uncertainty is very different from asserting "unpredictability" or declaring ignorance. Unpredictability is oftentimes used as an alibi to avoid difficult discussions, especially when dealing with long-term projects. At the other end of the spectrum, ignoring uncertainty and posturing more certainty than data or experience may justify is a sure way to lose credibility. The spectrum is not binary, not "black or white", nil or one. It is indeed a continuum from very high confidence to "guess work" where neither extreme position is certainly true.

At the end of the day, in spite of all the efforts, some people will remain dissatisfied, as the demand for certainty has shown itself to be a human trait since the time of Oracles, especially in times of crises. Taking the public's suggestions seriously into consideration will help reassure and give comfort with the additional benefit of perhaps shedding light on particular aspects of uncertainties. What remains most important is not offering things that cannot be done or achieved, or, again, misleading the public with simplistic assessments.

Oftentimes when explaining risks to management or the public the "zero risk" objection arises. Stakeholders may point out that any "non-zero" risk is unacceptable. In the aftermath of any recent accident, public opinion, regulators, law enforcement agencies and the media have vehemently embraced that vision. The first reaction should be to declare that stating a risk to be zero would be a lie due to inevitable uncertainties. We have been exposed to hazards and resulting risks since the dawn of humanity, but of course we are more ready to accept voluntary risks than involuntary ones. Nevertheless, it is a vain goal to expect zero risk in any human endeavor. More realistic goals may be "as low as reasonably possible" (ALARP) mitigation or the implementation of best available control technology (BACT) (see Sects. 13.2.1 and 14.3). Both these notions are of course open for discussion, as even their definitions are debatable and clouded with uncertainty (e.g., What is reasonable? What is possible? What is available?). The response to such questions depends on the assessment of the source and the quality of the request.

For operational risks, the demand for ALARP/BACT may be reasonable and lead to extremely low, tolerable exposures within sustainable mitigative budgets. In some cases the demand for zero risk may simply be a tactic to make a strong point. The demand then becomes a negotiating position. Here transparent tolerance criteria and uncertainties may help to explain that any further mitigation may divert capital from areas which require more attention.

Further down in our exploration, the demand for zero risk may be sincere but ill informed. This is when risk education becomes of the utmost important. The main topics for discussion may be, for example, zero risk does not exist; tolerance and acceptability; ALARA/ BACT definitions; and, of course, planned actions.

In some cases political motivations may also come into play. Opposing groups will oftentimes appeal to the community health awareness and base their approach on risks. Experts may be recruited to support this endeavor. In any case, attacking the legitimacy and sincerity of opposing groups is a poor approach, although it seems to have gained momentum in recent years, thanks to the relative impunity enjoyed by social media and the bad examples offered by politicians of all flavors and in many countries. One can disagree, but questioning the integrity of an opposing group or the right of the opposing group to use legitimate political strategies to promote their views is not a fair practice and will very likely backfire.

Finally, the demand for zero risk may be the result of distress stemming from outrage, anger, and distrust. That generally occurs together with negative judgments of the project promoter who becomes the enemy, e.g., the "ugly miner" focused on profits, arrogant, and dishonest. At that point risks are no longer the central issue. The root cause of hostile behavior may be a serious communication problem. Mending it is a labor of patience and tact: the goal is to understand what "broke the bridge", certainly not to try to convince that risks are bearable.

In (Oboni and Oboni 2016) we stated:

> The effects of today's risk mitigation programs will only slowly become visible.
>
> That is because the world-portfolio will contain mitigated and unmitigated (legacy) dams. During that time the public will perceive at best a status-quo.

We have read that the the owner of the Corrego do Feijao dam has reportedly been accused by Brazilian authorities of "not having learned anything from the prior failure." We are not in the position to discuss this point in detail due to lack of data. However, no one can expect a drastic change in the behavior of a large portfolio of dams, even if significant measures are taken. That is because, again, the root cause of the failures lies in decisions made decades ago, at the dams' inception, when the state of the art was different. Prioritizing actions on such a portfolio requires QRAs and is certainly not feasible with FMEAs and probability impact graphs.

The time factor is crucial in any risk communication, especially towards closure or during remediation, as was made apparent during the public hearings of several projects we have followed from a risk assessment point of view. In this respect, it may be worthwhile using more direct statements—for example, the expected number of accidents of a certain type over the life of the operation—rather than using the probability of failure.

References

Altarejos-García L, Silva-Tulla F, Escuder-Bueno I, Morales-Torres A (2015) Practical risk assessment for embankments, dams, and slopes, Risk and Reliability in Geotechnical Engineering, ed. Kok-Kwang Phoon & Jianye Ching, Chap. 11, Taylor & Francis

Bouder F, Slavin D, Loefstedt R (eds.) (2007) The Tolerability of Risk: A New Framework for Risk Management, Earthscan

Brehaut H (2017) Catastrophic dam failures path forward, Keynote lecture, Tailings and Mine Waste 2017, Banff, Nov 5–9, 2017

Caldwell J, Oboni F, Oboni C (2015) Tailings Facility Failures in 2014 and an Update on Failure Statistics, Tailings and Mine Waste 2015, Vancouver, Canada, October 25–28 2015 https://open.library.ubc.ca/media/download/pdf/59368/1.0320843/5

Chapman C, Ward S (2011) The Probability-impact grid - a tool that needs scrapping, in: How to manage Project Opportunity and Risk, 3rd ed., ch. 2, pp 49–51, Chichester, GB, John Wiley & Sons

Cox LA Jr, Babayev D, Huber W (2005) Some limitations of qualitative risk rating systems, Risk Analysis 25(3): 651–662

Espinoza, D (2015) The cost of risk: an alternative to risk adjusted discount rates, 10.13140/RG.2.1.4040.6643

Hance B, Chess C, Sandman P (1987) Improving Dialogue with Communities: A Risk Communication Manual for Government. Trenton, New Jersey, Office of Science and Research, New Jersey Department of Environmental Protection, December, 1987

Hubbard D (2009) Worse than Useless. The most popular risk assessment method and Why it doesn't work, in: The Failure of Risk Management. Ch. 7, pp. 117–144, Hoboken, Wiley & Sons

Kahneman D, Tversky A (1979). Prospect theory: An analysis of decision under risk. Econometrica, 47, 263–291.

[MVRB 2013] MacKenzie Valley Review Board (2013) Report of Environmental Assessment and Reasons for Decision Giant Mine Remediation Project http://reviewboard.ca/upload/project_document/EA0809-001_Giant_Report_of_Environmental_Assessment_June_20_2013.PDF

Oboni C, Oboni F (2013) Factual and Foreseeable Reliability of Tailings Dams and Nuclear Reactors -a Societal Acceptability Perspective, Tailings and Mine Waste 2013, Banff, AB, November 6 to 9, 2013

Oboni C, Oboni F (2018) Geoethical consensus building through independent risk assessments. Resources for Future Generations 2018 (RFG2018), Vancouver BC, June 16–21, 2018

Oboni F, Oboni C (2016) A systemic look at tailings dams failure process, Tailings and Mine Waste 2016, Keystone CO, USA, October 2–5, 2016 https://www.riskope.com/wp-content/uploads/2016/10/Paper-17_A-systemic-look-at-TD-failure-TMW-2016_07-11_revised.pdf

[Oboni et al. 2013] Oboni F, Oboni C, Zabolotniuk S (2013) Can We Stop Misrepresenting Reality to the Public?, CIM 2013, Toronto. https://www.riskope.com/wp-content/uploads/Can-We-Stop-Misrepresenting-Reality-to-the-Public.pdf

[Oboni and MDA 2018] Oboni Riskope Associates Inc., MacDonald Dettwiler and Associates, Ltd. (2018) Tailings 2.0 course: Space observation, quantitative risk assessment bring value and comply with societal demands, Tailings and Mine waste 2018, Keystone, Colorado

Robichek AA, Myers SC (1966) Conceptual Problems in the Use of Risk-Adjusted Discount Rates. The Journal of Finance, Vol. 21, No. 4 (Dec., 1966), pp. 727–730

Roche C, Thygesen K, Baker, E (eds.) (2017) Mine Tailings Storage: Safety Is No Accident. A UNEP Rapid Response Assessment. United Nations Environment Programme and GRID-Arendal, Nairobi and Arendal, ISBN: 978-82-7701-170-7 http://www.grida.no/publications/383

Silva F, Lambe TW, Marr WA (2008) Probability and Risk of Slope Failure, Journal of Geotechnical and Geoenvironmental Engineering 134(12) https://doi.org/10.1061/(ASCE)1090-0241(2008)134:12(1691).

Taguchi G (2014) Fault Tree Analysis of Slurry and Dewatered Tailings Management – A Frame Work, Master's thesis, University of British of Columbia

Chapter 16
One Final Word

This chapter includes some important notes on what the reader should be able to perform using this book.

Based on the steps discussed through this text and the references cited a reader should now be capable of starting a sustainable risk assessment for an operation or project and maintain it as long as needed. The goal has also been to enable the reader to:

- make recommendations to the board regarding the approval of corporate policies and standards for the management of material risks;
- oversee the policies and management processes used to manage the material sustainability risks of the corporation;
- review and assess assurance reports to verify that corporate policies and standards have, in fact, been implemented;
- review and assess management's risk management strategies for new projects and report to the board as to their adequacy;
- assess the application of adequate corporate resources;
- review and assess verification reports pertaining to the management of material risks.

If the reader is requested to develop a Tailings Operating Manual, leading practice would require:

- the identification through a Chap. 14-compliant methodology of risks with the highest level of attention being given to the development of appropriate critical control measures and procedures;
- the involvement, including final sign-off, of the engineer of record and other experts in the identification and preparation of critical control measures and procedures.

The Tailings Operating Manual should clearly identify the tolerable and manageable versus unmanageable risks and ensure that the greatest attention is paid to the development and documentation of critical control measures.

© Springer Nature Switzerland AG 2020   267
F. Oboni and C. Oboni, *Tailings Dam Management for the Twenty-First Century*,
https://doi.org/10.1007/978-3-030-19447-5_16

Mitigating strategies for operational and tactical risks should be identified for inclusion in the design and management system and to provide the basis for strong management control and regulatory oversight. Mitigation strategies that involve corporate policies and decision should be clearly pinpointed and described. However, it should be noted that such approval rests on assumptions related to the interpretation of guidances supporting the design process, the effectiveness of a company's governance system and the effectiveness of regulatory oversight. None of the above will be ethically valid if any kind of conflict of interest or complacency is in the way of the risk assessor.

The risk assessor should be neutral and directly or indirectly independent of the owners, promoters or engineers of the project/operation; he should report to the highest level of the board (i.e., the CEO and CFO and not to a sub-committee).

If the mining community is not capable of better evaluating and communicating risks or does not show pertinent actions and care SLO and CSR will not be fostered and decay.

It is time for mining companies, governmental agencies to benefit from better understanding the risks it is exposed to and exposes the public to.

Furthermore, we have to consider that:

- Unless proper methodologies are used it will be very difficult to evaluate progress, as factors such as climate change, seismicity and increase in population will further complicate the situation.
- Public outcry and hostility toward the mining industry, fueled by the diffusion of Information and Communication Technology will likely increase.
- As we stated in 2016, the effects of any risk mitigation program will only become visible over long-time spans, because any portfolio will contain mitigated and unmitigated (legacy) dams.
- During that time the public, regulators and legal authorities will perceive at best a status-quo, with obvious nefarious consequences to the owners and operators.

Chapter 17
Path Forward

This chapter defines what in our minds constitute the best possible evolution of world-class procedures.

The strategic intent for the mining industry must be to bring tailings dam failures and catastrophic incidents to the credibility threshold. The leading practices identified above, within the context of a comprehensive responsibility framework, will, if adopted, significantly contribute to a sustainable reduction of risks in tailings systems. However, such practices, although eliminating numerous fallacies of common practices, will still be afflicted by unpredictable natural events, judgment, human errors and varying levels of commitment.

For new tailings dams, risk-based rigorous deposition methods and site selection processes must be used to ensure that risks are corporately and societally tolerable. Existing operations must seek ways to reduce risks related to the original design. More importantly, companies, governments and geotechnical professionals must look to their own commitment and embrace the leading practices suggested in this text with the objective of providing the highest standard of risk management for their tailings dams.

In the short term, action should be taken to reduce the risks posed by using tailings impoundments to store water. For climates where water storage is inevitable, design requirements regarding freeboards, beach lengths and phreatic lines (water levels in the ground) should be clearly identified and strictly enforced by both corporate management and regulatory authorities. The use of tailings impoundments such as polishing ponds to reduce contaminant levels and to handle run-off and excess pit water must be stopped, as such practices only add to the risk level of a dam and tailings systems. Regulatory approvals for new TSFs must require the consideration of alternatives based on the minimization or elimination of water storage within the impoundment.

To support the consideration and adoption of lower consequence alternatives, a process must be initiated for the purpose of:

© Springer Nature Switzerland AG 2020
F. Oboni and C. Oboni, *Tailings Dam Management for the Twenty-First Century*,
https://doi.org/10.1007/978-3-030-19447-5_17

- bringing together the body of knowledge developed globally for each alternative deposition method;
- identifying and funding areas requiring further research or study; and
- developing technical guidances and leading practices for their design and operation.

For the longer term, it is difficult to identify what organization or what group of organizations will accept the challenge of moving aggressively towards the development of a comprehensive and leading-edge TRF. Unfortunately, this may have to wait for the next catastrophic failure. To prepare for that eventuality, useful progress could be made on focused priorities, which should be to:

- define the principles and standards of practice to be expected of a company's board of directors and provide appropriate protocols to guide and measure their implementation;
- prepare a guidance document to support the application of critical control methodology to the identification and management of critical tailings dam risks;
- define the principles, standards of practice and protocols required to guide an annual validation of the integrity of a tailings dam design, the adherence of the company to its regulatory and internal requirements and the implementation and maintenance of its critical control procedures;
- develop a model of a comprehensive and integrated assurance and reporting program that supports the needs of companies, governments and public.

Work must now start on the next round of incremental changes. We hope that this text will have dispelled any complacencies that lead to the belief the current situation is satisfactory and show the "way out" of common practices.

In fact, as this book was going to print, the tailings dam failure at Córrego do Feijão mine, Brumadinho (https://www.mindat.org/loc-145710.html), Minas Gerais (https://www.mindat.org/loc-387.html), Brazil took place, lending even greater urgency to the need to take the proper steps.

It is also hoped that many of our readers will work towards the further development and implementation of some of the ideas within their own organizations and the associations to which their organizations belong.

Finally, corporate tailings policy leading practice would require that the corporate tailings governance policy include commitments to:

- locate, design, construct, operate, and close tailings facilities in a manner that provides an acceptable level of protection for the safety, health, and welfare of the public and the environment;
- implement a tailings governance framework management system based the ISO 14001 environmental management standard or equivalent;
- utilize robust risk management systems and processes to identify and mitigate material risks;
- implement comprehensive change management and emergency preparedness and response plans;

- conduct an integrated risk-based tailings disposal method and site location selection process for new tailings dams based on a thorough understanding of the costs and consequences of failure of alternate methods and site locations;
- ensure the public is adequately informed of the nature of the risks relating to both pro-posed and existing tailings facilities and can effectively influence, in a collaborative manner, decisions that may interest or affect them;
- establish a comprehensive review and assurance program to verify that the commitments stated in the corporate governance policy are been met on a continuing basis and to provide the foundation for continual improvement; and
- make the assurance protocols and reports available to the public.

The achievement of the above objectives can be usefully informed by the methodologies explained in this book. We ardently hope that the suggestions and explanations set out here contribute to reducing risks and making our world a safer place.

Appendix A
Making Sense of Probabilities and Frequencies

Making sense of probabilities and frequencies (in a quantitative way) is necessary to benefit from better risk assessment. Oftentimes users feel compelled to use qualitative approaches to risk assessments. Their justification includes that probability evaluations are complicated, they require "statistics". As a result, users embrace index-approaches (probabilities are given (absurd) values like 1, 2, 3… n), qualitative approaches (small, medium, large… fast food style) while believing they will get a good understanding of their risks out of this.

This Appendix allows to simply define ranges of probabilities and frequencies based on analogies and tables, no "statistical" calculations required. By doing so, the user is forced to acknowledge uncertainties (resulting in expanding the estimate range) strongly reducing the "garbage-in-garbage-out" syndrome. Furthermore, if a consistent approach is used, then the relative estimates of probabilities will be consistent among each other.

A further benefit of using quantitative estimates is the ability to update those estimates as new data become available. Suppose you have selected ranges of min-max probabilities for all events. Those ranges of probabilities offer a way to use future observations to "correct" a priori estimates and thus build a dynamic risk register.

This makes sense, especially as we live in dynamic climate change, socio-economic, and political environments, and many companies use Internet of Things (IoT) or big data. Beware of the temptation to use a single number for estimates. Use ranges, as this is the way to acknowledge inevitable uncertainties.

© Springer Nature Switzerland AG 2020
F. Oboni and C. Oboni, *Tailings Dam Management for the Twenty-First Century*,
https://doi.org/10.1007/978-3-030-19447-5

A.1 Definitions

- **Frequency** is a measure of how often an event occurs on average during a unit of time (for example, how many times an engine supposed to start every morning fails to start per year). It ranges from 0 to infinite. Time can be replaced by other "counters". For people in charge of performing risk assessments a common unit of time is "per year". One can measure frequency by long term observations (building a "statistic").
- **Probability** is, by definition, a number between nil and one, measuring the chances some event may or may not happen. Nil means it is impossible to occur, one means it is certain to occur. If you want to stay out of jail never use nil or one!

A.2 Step 1
Making Sense of Probabilities and Frequencies

Table A.1 can be used for estimating a large probability, high frequency event x. For low probabilities $p_x \leq 0.1$, trained analysts can use Table A.2 delivered in Step 2, which zooms into the lower range of probabilities.

The last two columns to the right display the "Frequency equivalent of x" and the corresponding probability of seeing event x occurring (at least once) "next year". The next section delivers an explanation related to these two columns.

A.3 Step 2
Making Sense of Probabilities and Frequencies

Table A.2 starts where the prior Table A.1 finished (0.1) and reaches 10^{-6}, as this value is commonly considered to be the threshold value of human credibility. Going below credibility would require solid data that are normally not readily available, thus it is highly recommended to stay away from that range.

Here again, the last two columns to the right display the "Frequency equivalent" (return time) and the corresponding probability of seeing the event (at least once) "next year". The next section delivers an explanation related to these two columns.

Table A.1 Estimation of large probabilities

Colloquial vocabulary used to describe the event x occurrence	Event x	Frequency equivalent of x NB. If these events occur with a known average rate and independently of the time since the last event	P_x to see at least one event next year p_{xmin}–p_{xmax}
Usually, Almost always	Finding at least one container of ice cream in a family freezer	≥ 1	0.63 to ~1.0
	At least one sunny week-end in the next year		
Common, Must be considered, Not always	A member of the family gets a cold next year	0.7–1	0.5–0.63
	Getting stuck in a traffic jam for at least 20 min next year (exclude commuting)		
Not uncommon	A person between the age of 18 and 29 does NOT read a newspaper regularly	0.36–0.7	0.3–0.5
May be, Possibly	Getting stuck for more than one hour in traffic (exclude commuting). A celebrity marriage will last a lifetime	0.23–0.36	0.2–0.3
Not usually, Occasionally	Odds of dying from heart disease or cancer in the US (1/7) chance of drawing 1 when drawing a fair dice (1/6 = 0.16)	0.11–0.23	0.1–0.2
Rarely Almost never Never	*NB: a non expert should stop at this level of scrutiny.* Experts can develop more in depth estimates for lower probabilities levels using the next Table A.2		0–0.1

A.4 Explanation Related to the Last Two Columns to the Right

For small frequencies (up to 1/10 yrs (units)), one can assume annual probability = frequency, as shown in Fig. A.1.

However, at frequency = 1/5 yrs (f = 0.2) the error of the approximation rises to 20%. After that, calculations are required to evaluate the probability if the frequency is known or vice versa.

For those who want to know a bit more, frequencies and probabilities are indeed linked by a mathematical function known as the Poisson distribution. It is a common mistake to believe that a frequency f means that the event will occur "once every 1/f year" as that event can indeed occur 0, 1, 2, ... n times with decreasing probability within a selected interval. As a matter of fact, the probability of an event occurring once or more within its return period is always 0.63.

Table A.2 Lower range probabilities

Likelihood of "rare" phenomena	Event x	Return time (years) prob = Frequency equivalent = 1/return time	P_x to see the even next year $p_{x\ min} - p_{x\ max}$
High	Having an income of more than 700k US$ in the US in 2017 (1 in 100)/higher bound of likelihood to have a 7.0 or even higher magnitude on the San Andreas Fault line	100–10	0.01–0.1 $(10^{-2}-10^{-1})$
Moderate	Assault by firearm in the US (237 in 100,000 inhabitants)	1000–100	0.001–0.01 $(10^{-3}-10^{-2})$
Low	Influenza death (1 in 5000 to 1 in 1000) per person/an earth tailings dam breaches on earth per year	10,000–1000	0.0001–0.001 $(10^{-4}-10^{-3})$
Very low	Fatal accident at work (1 in 43,500 to 1 in 23,000) per worker/class 5+ nuclear accident on earth	100,000–10,000	0.00001–0.0001 $(10^{-5}-10^{-4})$
Extremely low	Person stricken by lightning (1 in 161,856)	1,000,000–100,000	0.000001–0.00001 $(10^{-6}-10^{-5})$
Credibility threshold lower likelihoods exist	Fatality in railway accident (travelling in Europe) (0.15 per Billion km)/meteor landing precisely on your house; a major Swiss hydro-dam breaching	N/A	Unless data abound, lower values should not be used

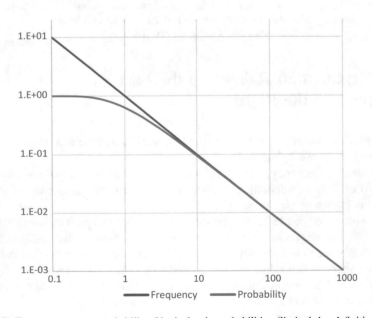

Fig. A.1 Frequency versus probability. Vertical axis probabilities (limited, by definition, to 1), horizontal axis corresponding value of 1/frequency. Frequency expressed in events per year

Further Reading

[ANCOLD 2012] Australian National Committee on Large Dams (2012) Guidelines on Tailings Dams – Planning, Design, Construction, Operation and Closure. https://www.ancold.org.au/?page_id=334

[AG CMH 2016] Australian Government (2016) Cyanide Management. https://www.industry.gov.au/sites/default/files/2019-04/lpsdp-cyanide-management-handbook-english.pdf

[BC AMP 2016] BC Mines and Mineral Resources Division (2016) Administrative Monetary Penalties Discussion Paper. https://www2.gov.bc.ca/assets/gov/farming-natural-resources-and-industry/mineral-exploration-mining/documents/compliance-and-enforcement/amp_discussion_paper_september2016.pdf

[BC C&E 2016] BC Government (2016) Deputy Ministers Mining Compliance and Enforcement (C&E) Board. http://www2.gov.bc.ca/gov/content/industry/mineral-exploration-mining/compliance-enforcement/board

[BC Code 2017] BC Government (2017) Health, Safety and Reclamation Code for Mines in British Columbia. BC Ministry of Energy and Mines. http://www2.gov.bc.ca/gov/content/industry/mineral-exploration-mining/health-safety/health-safety-and-reclamation-code-for-mines-in-british-columbia

[BC Info 2107]. BC Mine Information website. http://mines.nrs.gov.bc.ca/

Golder Associates. 2016. *Review of tailings management guidelines and Recommendations for improvement.* http://www.icmm.com/tailings-report

[ICMM]. International Council on Mining & Metals website. http://www.icmm.com/en-gb/about-us

[ICMM 2005] International Council on Mining & Metals (2005) Good practice in emergency preparedness and response. ICMM & UNEP http://www.icmm.com/website/publications/pdfs/health-and-safety/good-practice-emergency-preparedness-and-response

[ICMM 2015] International Council on Mining & Metals (2015) Health and safety critical control management: good practice guide http://www.icmm.com/en-gb/environment/tailings

Institute For Systems Informatics And Safety, Guidance On Land Use Planning As Required By Council Directive 96/82/Ec (SEVESO II), ISBN 92-828-5899-5, (Christou, M.D., Porter, S., EUR 18695), European Communities, 1999

[King III 2009]. Institute of Directors in Southern Africa (2009) King Report on Governance for South Africa http://www.iodsa.co.za/?kingIII

[MAC Audit 2011]. Mining Association of Canada. A Guide to Audit and Assessment of Tailings Facility Management. MAC. http://mining.ca/towards-sustainable-mining/protocols-frameworks/tailings-management

© Springer Nature Switzerland AG 2020
F. Oboni and C. Oboni, *Tailings Dam Management for the Twenty-First Century*,
https://doi.org/10.1007/978-3-030-19447-5

[MAC Guide 2011]. Mining Association of Canada (2011) The Guide to the Management of Tailings Facilities. http://mining.ca/towards-sustainable-mining/protocols-frameworks/tailings-management

[MAC Manual 2013]. Mining Association of Canada (2013) Developing an Operation, Maintenance and Surveillance Manual (OMS Manual) for Tailings and Water Management Facilities. MAC. http://mining.ca/documents/developing-operation-maintenance-and-surveillance-manual-tailings-and-water-management

[MAC OMS Guide 2019] Mining Association of Canada (2019) Developing an Operation, Maintenance, and Surveillance Manual for Tailings and Water Management Facilities SECOND EDITION, Feb. 2019

[MAC TRF 2015]. Mining Association of Canada (2015) Report of the TSM Tailings Review Task Force. MAC. http://mining.ca/towards-sustainable-mining/protocols-frameworks/tailings-management

[WA 2013] Government of Western Australia (2013) Code of practice: Tailings storage facilities in Western Australia. http://www.dmp.wa.gov.au/Documents/Safety/MSH_COP_TailingsStorageFacilities.pdf

Wilson, G. Ward & Robertson, Andrew MacG. 2015. *The Value of Failure.* Geotechnical News, June 2015

Working Group on Landslides, Committee on Risk Assessment, Quantitative Risk Assessment for Slopes and Landslides: The State of the Art, IUGS Proceedings, Honolulu, Balkema, 1997